AF678889

Ship Sensors

When combined with artificial intelligence, advanced computing architectures and enhanced communications, sensor technologies can monitor vessel performance and the adjacent environment to detect conditions that may hinder voyage completion. This book provides insight into the present and future of sensor architectures and configurations that can enhance vessel performance and further improve the safety of navigation. It covers topics such as traditional and expanded sensor functions in engineering and navigation, as well as new sensor capabilities that can provide greater insight into vessel behavior and performance and enhance awareness of passenger, crew and other human activities.

Chapters offer background information on typical legacy vessel sensor configurations and current International Maritime Organization (IMO) requirements for onboard sensors and future regulatory trends before discussing modern sensors and current-generation "smart" sensors that provide enhanced situational awareness to watchstanders. A vision of next-generation sensors currently being investigated for shipboard use is provided along with long-term trends in quantum sensing and computing that promise radical change across a wide variety of vessel functions. Insight is also given into cybersecurity factors so essential to all sensor systems.

Ship Sensors: Conventional, Unmanned and Autonomous is ideal for professional seafarers, maritime academics and university students, and developers of maritime sensors and systems.

R. Glenn Wright holds a BS in Electrical Engineering, MS in Computer Science and PhD in Maritime Affairs with 45 years of industry experience leading research in sensor-based systems for surface vessels and unmanned underwater vehicles, meteorological and oceanographic data systems, and the automated testing of electronic equipment. He is also a master mariner and research vessel captain investigating meteorological, oceanographic, and electromagnetic phenomena and sensor data fusion, with direct knowledge of ship operations and the application and usefulness of the foundational technologies involved. This book complements his book *Unmanned and Autonomous Ships*, also published by Routledge.

Ship Sensors

Conventional, Unmanned and Autonomous

R. Glenn Wright

Routledge
Taylor & Francis Group
NEW YORK AND LONDON

Cover image: © Johnny Haglund/Getty Images

First published 2024
by Routledge
605 Third Avenue, New York, NY 10158

and by Routledge
4 Park Square, Milton Park, Abingdon, Oxon, OX14 4RN

Routledge is an imprint of the Taylor & Francis Group, an informa business

© 2024 R. Glenn Wright

The right of R. Glenn Wright to be identified as author of this work has been asserted in accordance with sections 77 and 78 of the Copyright, Designs and Patents Act 1988.

All rights reserved. No part of this book may be reprinted or reproduced or utilised in any form or by any electronic, mechanical, or other means, now known or hereafter invented, including photocopying and recording, or in any information storage or retrieval system, without permission in writing from the publishers.

Trademark notice: Product or corporate names may be trademarks or registered trademarks, and are used only for identification and explanation without intent to infringe.

Library of Congress Cataloging-in-Publication Data
Names: Wright, R. Glenn, author.
Title: Ship sensors : conventional, unmanned and autonomous / R. Glenn Wright.
Description: New York, NY : Routledge, 2024. |
Includes bibliographical references and index.
Identifiers: LCCN 2023042461 | ISBN 9781032456218 (hbk) |
ISBN 9781032456225 (pbk) | ISBN 9781003377900 (ebk)
Subjects: LCSH: Ships–Electronic equipment. |
Intelligent sensors. | Sensor networks.
Classification: LCC VM480 .W75 2024 |
DDC 623.8/504–dc23/eng/20231108
LC record available at https://lccn.loc.gov/2023042461

ISBN: 978-1-032-45621-8 (hbk)
ISBN: 978-1-032-45622-5 (pbk)
ISBN: 978-1-003-37790-0 (ebk)

DOI: 10.1201/9781003377900

Typeset in Sabon
by Newgen Publishing UK

Contents

Figures

Tables

Preface

Ship sensors are being viewed from a new perspective that considers the fusion of their data, capabilities and performance through a comprehensive approach that sees every aspect as a cohesive entity rather than a collection of isolated sensors and systems. Driven by revolutionary technological advances by vessel autonomy initiatives, new sensor capabilities can enhance seafarer awareness of vessel status, identify trends in vessel performance and the adjacent environment that improve potential for voyage success and detect conditions that hinder and may jeopardize successful voyage completion. Combined with artificial intelligence, advanced computing architectures and enhanced communications, the management and control of shipboard systems can achieve broad new capabilities to improve safety for traditionally staffed vessels and implement new strategies to achieve full independence for autonomous vehicles.

This book on ship sensors logically follows the author's previous book, *Unmanned and Autonomous Ships*, providing a detailed look at the various organs and appendages comprising a vessel's architecture that makes possible partial and even full automation of various onboard functions. However, sensors are increasingly prevalent throughout all levels and functions of conventionally staffed ships, making this topic vitally relevant for all vessels. The ultimate goal is to provide a window into the present and future of sensor architectures and configurations that can enhance vessel performance, safety and further improve the safety of navigation. In an attempt to eliminate potential confusion, unmanned and autonomous watercraft of all types and sizes are referred to as "vehicles" to distinguish them from the traditional use of the word "ships" throughout the chapters of this book. The word "vessels" can apply to both ships and vehicles.

Topics covered include traditional and expanded sensor functions in engineering and navigation as well as new sensor capabilities integrated into occupied ship's spaces and cargos that can provide greater insight into vessel behavior and performance and enhance awareness of passenger, crew and other human activities. Background is provided regarding typical legacy vessel sensor configurations and current IMO requirements for onboard sensors and future regulatory trends. Discussed also are modern sensors and current-generation "smart" sensors that provide enhanced situational awareness to watchstanders. A vision of next-generation sensors currently being investigated for shipboard use is provided along with long-term trends in quantum sensing and computing that promise radical, revolutionary and game-changing alterations across a wide variety of vessel functions. A topic unique to this discussion is that of sensor degradation in terms of what it is, how it may be detected and the potential for overcoming the many limitations that result from its occurrence.

Additional topics include the means by which sensor data communication is accomplished and how data fusion can achieve far greater promise to enhance situational awareness and safety of navigation than the sum of the individual sensors themselves. Insight is provided concerning cybersecurity essential to all sensor systems, the means by which sensor data can be protected and how confidence can be assured in the information they provide across an operational environment. Also discussed is how sensor system design can be verified with respect to the requirements they are intended to fulfill, validated to ensure they actually accomplish these requirements and tested to determine their resilience and robustness under a wide variety of conditions.

Two definitions are offered to help provide context for the discussions in this book:

Automaton[1] (noun) au· tom· a· ton

1 A mechanism that is relatively self-operating, especially a robot.
2 A machine or control mechanism designed to follow automatically a predetermined sequence of operations or respond to encoded instructions.

Smart Sensor[2]

1 A sensor that conditions measured signals before transmitting them to the control network.

Notes

1 www.merriam-webster.com/dictionary/automaton
2 www.pcmag.com/encyclopedia/term/smart-sensor

Acknowledgments

My eternal love, thanks and appreciation are extended to Magdalena, whose ability to achieve software miracles and results that make this and all of our work possible is exceeded only by her inspiration and support throughout these endeavors. I am forever most grateful.

My thanks to Adam Zgol who greatly aided in performing the experiments used in detecting sensor degradation described in Chapter 10, and who was also instrumental in establishing the underwater remotely operated vehicle (ROV) capabilities used in our research.

Passing mention is made in the way of thanks to my two freelance editors, Rubi (English Bulldog) and Semper (German Shepherd), who drooled on the pages of this manuscript they liked and ripped up those they did not like. Far too often they were correct.

Acronyms and Abbreviations

ABS	American Bureau of Shipping
AC	Alternating Current
A/D	Analog to Digital (Conversion)
AI	Artificial Intelligence
AIS	Automatic Identification System
AM	Amplitude Modulation
APT	Automatic Picture Transmission
ARPA	Automatic Radar Plotting Aid
AtoN	Aid to Navigation
BeiDou	Navigation Satellite System (China)
BIMCO	Baltic and International Maritime Council
CAN	Controller Area Network
CCS	Cargo Control System
CNN	Convolutional Neural Network
COLREGS	Convention on the International Regulations for Preventing Collisions at Sea
COSPAS	Space System for Search of Distress Vessels (Russia)
COTS	Commercial Off The Shelf
D/A	Digital to Analog (Conversion)
DARPA	Defense Advanced Research Projects Agency (United States)
DC	Direct Current
DCSA	Digital Container Shipping Association
DGPS	Differential Global Positioning System
DoD	Department of Defense (United States)
DoS	Denial of Service
DR	Dead Reconning
DSC	Digital Select Calling
ECDIS	Electronic Chart Display Information System
ECU	Electronic Control Unit
eLORAN	Enhanced Long-Range Radio Navigation
EHF	Extremely High Frequency

ENC	Electronic Navigation Chart
EPIRB	Emergency Position Indicating Radiobeacon
FM	Frequency Modulation
FLIR	Forward-looking Infrared
FR	Fouling Rating
Galileo	Global Navigation Satellite System (European Union)
GAN	Generative Adversarial Network
GHz	Gigahertz
GLA	General Lighthouse Authorities of the United Kingdom and Ireland
GMDS	Global Maritime Distress and Safety System
GNSS	Global Navigation Satellite System
GPS	Global Positioning System (United States)
GT	Gross Tons
HEO	High Earth Orbit
HF	High Frequency
HVAC	Heating, Ventilation and Air Conditioning
Hz	Hertz (frequency)
IACS	International Association of Classification Societies
IALA	International Association of Marine Aids to Navigation and Lighthouse Authorities
IAPH	International Association of Ports and Harbors
IBS	Integrated Bridge System
IC	Integrated Circuit
IMO	International Maritime Organization
INS	Inertial Navigation System, Integrated Navigation System
IoT	Internet of Things
IR	Infrared
ISC	International Code of Signals
ISO	International Organization for Standardization
IT	Information Technology
kHz	Kilohertz
LAN	Local Area Network
LDUUV	Large Displacement Unmanned Undersea Vehicle
LEO	Low Earth Orbit
LF	Low Frequency
LIDAR	Light Imaging Detection and Ranging
LORAN	Long-Range Radio Navigation
LUSV	Large Unmanned Surface Vehicle
mA	Milliamperes
MAIB	Marine Accident Investigation Branch (UK)
MASS	Maritime Autonomous Surface Ships

MAtoN	Mobile Aid to Navigation
MEO	Medium Earth Orbit
MF	Medium Frequency
MHz	Megahertz
mmRADAR	millimeter RADAR
MMSI	Maritime Mobile Service Identity
MSC	Maritime Safety Committee
MSI	Maritime Safety Information
MTS	Maritime Transportation System
MUSV	Medium Unmanned Surface Vehicle
NavIC	Navigation Indian Constellation (India)
NAVTEX	Navigational Telex
NIST	National Institute of Standards and Technology (United States)
NMHS	National Meteorological & Hydrological Services
NTSB	National Transportation Safety Board (United States)
NTSC	National Science and Technology Council (United States)
ODU	Offshore Drilling Units
OSI	Open Systems Interconnection
OT	Operational Technology
PC	Personal Computer
PCS	Port Control System
PLA	Programmable Logic Array
PLC	Programmable Logic Controller
PM	Pulse Modulation
PNT	Positioning, Navigation and Timing
PRM	Parametric Rolling Movement
QZSS	Quasi-Zenith Satellite System (Japan)
RACON	Radar Beacon
RADAR	Radio Detection and Ranging
RDF	Radio Direction Finding
RDP	Remote Desktop Protocol
RF	Radio Frequency
RFID	Radio Frequency Identification
ROC	Remote Operations Center
Ro-Ro	Roll on-Roll off
RTD	Resistance Temperature Detector
SAIS	Satellite-AIS
SAR	Synthetic Aperture Radar, Search and Rescue
SARSAT	Search and Rescue Satellite-Aided Tracking
SART	Search and Rescue Radio Transmitter
SAS	Synthetic Aperture Sonar

SBAS	Satellite-Based Augmentation System
SOLAS	Safety of Life at Sea convention
SSB	Single Side Band
TCP/IP	Transmission Control Protocol/Internet Protocol
TEU	Twenty-Foot Equivalent Unit
THz	Terahertz
TTE	Time-Triggered Ethernet
μA	microamperes
UMAA	Unmanned Maritime Autonomy Architecture
U.S.	United States of America
USV	Unmanned Surface Vehicle
UUV	Unmanned Undersea Vehicle
VATON	Virtual Aid to Navigation
VHF	Very High Frequency
VHF-ATON	VHF Aid to Navigation
VLCC	Very Large Crude (oil) Carrier
VLF	Very Low Frequency
VPN	Virtual Private Network
VSAT	Very Small Aperture Terminal
VTMIS	Vessel Traffic Management Services
VTS	Vessel Traffic Services
WAN	Wide Area Network
Wi-Fi	*Registered trademark of the Wi-Fi Alliance*
WIG	Wing in Ground craft
WMO	World Maritime Organization, World Meteorological Organization
XLUUV	Extra Large Unmanned Undersea Vehicle

Chapter 1

Introduction to Ship Sensors

This book describes a ship's senses of vision, hearing, smell, touch and taste, which are vital to how a vessel and her crew, human and cyber, interpret and react to things that are going on within and to the external world outside the hull, superstructure and on deck. Events taking place may aid or hinder the goal of planning, undertaking and completing a voyage. Watchstanders on the bridge ensure the ship can make way and doesn't bump into things en route. Crewmembers in the engineering department are entrusted with propelling the ship through the water as well as operating and maintaining the various onboard systems that make the ship come to life. Deckhands ensure the safe and efficient loading, overseeing, and unloading of cargos and any passengers that may be along for the ride; except when the number of passengers begins to exceed crew sizes, where hospitality staff are entrusted with the duties of passenger safekeeping and morale.

To illustrate the roles played by human beings and ship sensors, two short stories depicting different points of view of the same sequence of events are told. The first describes an observer who assigns their own thoughts and feelings to events as they unfold. The second describes the underlying processes to interpret these events and make appropriate decisions and take action. One is very human, the other a matter of fact.

Ship Without A Soul

Standing at the rail in the starlit night and listening intently to the silence, the sea was glazed with skim ice that appeared in the early hours of the morning. The moon was quartered and shrouded in mystery. To starboard the fog shimmered into the air, to port the reflected moonlight reached to the horizon. No wind, no movement, adrift, awaiting our turn to proceed further.

Then came the sound. Clinking and tinkling like a thousand champaign glasses. There was a growing sense of anticipation as the echoes grew louder, yet a pleasantness filled the air. You could feel her presence. A soft

DOI: 10.1201/9781003377900-1

glow appeared in the fog and gradually her sleek lines were revealed. White, then red shown through the silver apparition.

I knew her name, where she was going and where she was from.

She moved with purpose intent on making her way, knowing I was there but she would not stay.

Cold and indifferent, no crew is aboard. Emotionlessness without feelings, but nothing untoward.

Neither a ghost from the past, nor anything old. Just an automaton of today. A ship without a soul.

This, of course, is a story of two ships passing in the night except with a modern twist where one vessel is unmanned and autonomous, operating and navigating using her own resources. The ship's presence and identity are announced to others using Automated Identification System (AIS), with her every move choreographed by an automated captain viewing the world through electronic eyes, ears and other senses based on the knowledge of what has been learned from terabytes of training data. However, none of this is possible without conventional and unmanned vessels being laden with sensors to provide full-time situational awareness across all engineering functions throughout the ship and navigation functions below, on and above the sea during the voyage. Both vessels in this story have benefitted from advances in ship automation. It is hoped that the contents of this book help to bring these concepts into view in a logical manner.

1.1 Ship Sensors

The use of ship sensors has long been considered in the context of the engineering and deck departments. However, new technologies now also consider sensor applications that have expanded into the sensing of crew and passenger activity and cargo characteristics, new vessel systems and operational parameters, and many other factors previously incapable of being observed. Smart sensor technologies that provide far greater insight into system performance and trends have also blurred these traditional areas of responsibility as interrelationships between factors having to do with engineering, navigation, deck, passenger and cargo operations become more complex. Related concepts are also introduced in subsequent chapters to stimulate discussion on sensor data communications, cybersecurity of sensor applications and communications, sensor degradation, and future trends in sensor design and operation.

1.2 What Exactly Is a Sensor?

Simply stated, a sensor may be characterized as a device that detects and responds (usually with an electrical signal) to an input from the physical environment.[1] In practice, the terms "sensor" and "transducer" are often confused, but for the sake of discussion, they are interchangeable with each other. In the context of ships and unmanned vehicles, these inputs can come from on or within the vessel itself, as in the case for engineering, passenger, and cargo sensors; or external to the vessel from the surrounding sea, air and space as may heighten situational awareness of engineering and bridge watchstanders as well as autonomous agents tasked with advising seafarers and in decision-making roles to increase safety of navigation. These sensors are typically connected to electronic displays in engineering spaces and on the bridge, to internal networks that distribute sensor data and information derived from sensor data to locations within the vessel for fusion, additional processing and sensor analytics. Sensor data and information can also be distributed external to the vessel through broadband communication channels and via "the Cloud" to remote centers worldwide for vessel monitoring and control. The analysis of diverse data from multiple sensors using different sensing methods can provide greater insight and dimensionality into behavior and processes to formulate situational awareness and problem solutions than the individual sensors themselves, with the whole picture being greater than the sum of the individual inputs. An introduction is provided to several of these topics in the following paragraphs.

1.3 The Sensor Revolution

New sensors and innovations in their application and use are expanding our insight into not only the world and the universe that surrounds us, but also our perception and knowledge of how things operate and the processes they use. Accompanying these sensors has been the evolution of computer hardware technology that facilitates the rapid processing of great volumes of sensor data. Statistical analysis and artificial intelligence-based software make possible the detection of trends, attributes and objects based on the characteristics of sensor data contents and enable the subsequent taking of appropriate action.

One example of how sensor technology has evolved is the breathtaking imagery obtained in 1995 by the U.S. National Atmospheric and Space Administration (NASA) Hubble Space Telescope of the "Pillars of Creation" located some 6,500–7,000 light-years from Earth revealing never before seen details of three giant dark columns of cold gas. The Hubble telescope represented the most advanced technology of its era. Yet in less than 30 years the James Webb Space Telescope in 2022, using newly developed near-infrared imaging technology, can now see inside these columns to reveal internal structures filled with many stars and interactions in the process of star formation.[2] During this same period enormous advances were made in medical imaging technologies that include ultrasound, magnetic resonance imaging (MRI) and computed tomography (CT) that, through the use of modern digital computing, today makes possible non-invasive visualization of the interior human body with unprecedented detail and resolution.[3] Simultaneously, at the small end of the size spectrum scientists with the aid of atomic force microscopy (AFM) are now able to image, measure and manipulate particles as small as 10 nano-meters (nm) or 0.000000001 meters (m) in size for the development of new medicines.[4]

Nestled snugly between galactic and atomic dimensionalities lie ship sensors and sensor technology that have also continued to evolve and change the face of shipping and ships as we have traditionally known them. Significant advances have been made across all aspects of ship sensors in terms of monitoring the functions and processes of the various onboard engineering, navigation, cargo, and passenger systems which are the subjects of this book.

1.4 Non-technical Factors Driving Maritime Sensor System Innovation

Much of the innovation in shipboard sensor systems stems from industry and International Maritime Organization (IMO) digital ship and related initiatives designed to accommodate advances in artificial intelligence and smart sensors with greater data content, volume and flow, and to integrate these new and advancing technologies into the regulatory framework.[5] These include developments associated with Maritime Autonomous Surface Ships (MASS) and generally smaller Unmanned Surface Vehicles (USV); and issues pertaining to enhanced Safety of Life at Sea (SOLAS), modernization of the Global Maritime Distress and Safety System (GMDSS), e-Navigation and resulting enhancements and improved maritime services, and the standardization of shipborne satellite navigation system receiver equipment.

1.5 Electromagnetic Spectrum Allocation to Maritime Sensors

Any discussion on ship sensors is incomplete without considering the characteristics of the physical environment in which they operate. Using human senses and our many sensory limitations as a baseline, our visibility into the world must be considered deficient at best. Many birds and cats see clearer and at greater distances, especially at night. Dogs have a much better sense of smell, and wolves can hear over great distances. Whales and elephants can hear the lowest frequencies; the Great Wax moth has the highest ultrasound hearing, while the rabbit has the best overall hearing range amongst land animals. Insects such as beetles have no hearing at all but make up for this by being sensitive to vibrations through a sense of touch. However, human beings have an immense ability to observe other creatures and adopt their unique skills for themselves. Through observation and experimentation, we have created sensors that provide us with awareness where we can communicate great volumes of related information across large portions of the electromagnetic spectrum, thereby appropriating many common capabilities from other inhabitants of this planet. Table 1.1 illustrates the scope of many sensors available today, along with their operating and data communication frequencies in the electromagnetic spectrum.

Note that there exist many references throughout the literature to different band and sub-band allocations within the Extremely High Frequency (EHF) millimeter-wave frequencies, including Q, U, E, F, D and G-band designations not identified in Table 1.[6,7] Considering ongoing levels of research and experimentation with these frequencies and for clarity, we identify mmRadar as operating in the W-band.

Immediately evident is the fact that seafarers can now have super-human capabilities to sense objects and events. Indeed, distance is no longer a limitation as we also have the ability to communicate sensor data and related information in real time to any location on Earth. However, gaining new powers to see, hear and communicate provides little advantage without adequate training in their use. Indeed, a lack of knowledge can often lead to overconfidence in our decision-making capabilities with disastrous results, especially when we are unaware of the limitations of these new senses.

The evolution of all types of ship engineering, navigation, cargo, and human activity sensors since 1995 has been remarkable, with increases in speed, clarity, resolution, accuracy and utility occurring seemingly almost on a daily basis. The following paragraphs elaborate on these developments as they presently exist in the maritime world and sheds some light on many of the near-future trends that continue to revolutionize the industry and dramatically affect the lives of seafarers worldwide.

1.6 Types of Ships and Maritime Vehicles

Before describing sensors used onboard ships, it is necessary to provide context for their use in the many different types of ships on which they reside and the specialized roles they fulfill in world commerce. A partial list of the types of ships considered in this book includes:

- Passenger Ships
- Container Ships
- Tanker Ships
- Bulk Carriers
- Gas Carriers
- General Cargo
- Dry Cargo
- Refrigerated Cargo
- Ro-Ro Ships
- Livestock Carriers
- Offshore Ships
- Fishing Vessels

Table 1.1 Electromagnetic Spectrum Allocation to Maritime Sensors

	Frequency Range	*Frequency Units*	*Sub-Band*		*Maritime Navigation Sensors*
RADIO	30–300 kHz	Low Frequency (LF)	**Human hearing: 20 Hz–20 kHz**		Sonar, Acoustic underwater communication, Engineering, eLoran
	300 kHz–3 MHz	Medium Frequency (MF)			MF SSB, GMDSS, Sonar, RDF, NAVTEX, Engineering
	3–30 MHz	High Frequency (HF)			HF SSB communications, GMDSS, RDF
	30–300 MHz	Very-High Frequency (VHF)			VHF communications, EPIRB, VTS, GMDSS, AIS, RDF, APT
			200–250 MHz	G-band	Few maritime services, possibly Radar
			250–500 MHz	P-band	EPIRB Satellite Signal, SAR
	300 MHz–3 GHz	Ultra-High Frequency (UHF)	500–1,500 MHz	L-band	GNSS, Mobile Phone, Inmarsat, Weather, WAN, Cargo, EPIRB
			2–4 GHz	S-band	GNSS, Satellites, Weather, Microwave, Radars, LAN, Bluetooth
	3–30 GHz	Super-High Frequency (SHF)	4–8 GHz	C-band	Communication and Weather satellites, Wireless, Cargo
			8–12 GHz	X-band	Meteorological, Earth Observation/Comm. Sats, Radar, SART
			12–18 GHz	Ku-band	Communication and Weather satellites
			18–27 GHz	K-band	Communication satellites, mmRadar

	30–300 GHz	Extreme-High Frequency (EHF)	27–40 GHz	Ka-band	Communication and Weather satellites
			40–75 GHz	V-band	Radars, Communication satellites
			75–111 GHz	W-band	mmRadar
OPTICAL	300 GHz–400 THz	Infrared Light (IR)			Infrared Sensors and Cameras, Lidar
	400–750 THz	Human Visible Light (VIS)			Cameras

Source:

ITU 2015. Nomenclature of the frequency and wavelength bands used in telecommunications. Recommendation ITU-R V.431-8 (08/2015). International Telecommunications Union, Geneva.

These exemplify the vast majority of sensor applications found on modern and future vessels most likely to be encountered by seafarers and outfitters. General discussions about common sensor applications across these and other types of vessels are included in the following paragraphs throughout this chapter.

Naval ships feature many commercially available engineering and bridge navigation sensors discussed in this book, which are common among all ships. However, no attempt is made to address tactical information nor naval-specific sensors such as acoustic, Sonar, Radar, and shipboard air and torpedo defense systems unique to operating within the maritime battlespace for which technical specifications are not readily available.

The text also does not address special-purpose vessels with specific and unique engineering requirements and sensors beyond those common to most ships. There are many in this classification, including nuclear and wind-powered ships, submarines, surface effect ships, hovercraft, hydrofoils, wing-in-ground (WIG) craft, liftboats, offshore drilling units (ODU), ice breakers, cable and pipe layers, research vessels, tugboats, barge and timber carriers, salvage vessels, diver support ships, survey vessels, buoy tenders, crane barges, dredgers, floating cranes and other such vessels.

1.7 Sensors Discussed, and Not Discussed

The heart of this book discusses sensors associated with monitoring engineering, navigation, cargo, and human functions onboard ships. However, not covered and acknowledgment is made regarding numerous sensors found onboard ships but also commonly found across many maritime and land-side applications. These include sensors associated with the operation and monitoring of lifts and elevators, housekeeping and food services, cooking facilities, swimming pool and spa machinery, entertainment systems and other functions that exist on ships, from the smallest of vessels to large-scale and complex operations that rival the world's largest hotels, restaurants and catering establishments.

1.8 Flow of the Chapters in This Book

The chapters of this book are organized to introduce many of the essential elements of shipboard sensor systems, provide specifics on several of their key features and then touch on some of the peripheral subjects that directly affect sensor operations. This is followed by discussion of the author's perception of future near and long-term trends in sensor design, technology and applications.

Chapters 2–10 focus on the present state of sensor technology, describe many types of sensors likely to be found in and on various modern ships and unmanned vehicles and highlight some important sensor-related issues being considered throughout the industry.

Chapters 2, 3, 4 and 5 discuss sensors from perspectives specific to functions associated with engineering, navigation, cargo, crewmembers and passengers, respectively.

Chapter 6 provides background into essential technologies associated with artificial intelligence (AI) and how these are revolutionizing sensor designs by creating smart sensors and sensor analytics that provide unprecedented insight into sensed phenomena and processes. Also discussed is the use of communications as a means to extend human senses worldwide.

Chapter 7 identifies many of the methods by which sensor data is communicated between ships and from ship to shore.

Chapter 8 describes how sensor data is communicated within the ship itself.

Chapter 9 addresses the critical subject of sensor system cybersecurity and how this fits in with overall vessel cybersecurity.

Chapter 10 addresses the subject of sensor degradation that stymies sensor operation and performance and how this phenomenon can adversely affect vessels of all types in voyage completion, as well as introduce factors that decrease the safety of navigation and

Chapter 11 bridges the gap between the previous chapters, future enabling technologies, and the directions sensors will likely to take regarding their application and use.

References

1 Robert Sheldon. What Is a Sensor? WhatIs.com, TechTarget. August 2022. www.techtarget.com/whatis/ definition/sensor

2 Tereza Pultarova. Why the James Webb Space Telescope's Amazing 'Pillars of Creation' Photo Has Astronomers Buzzing. Space.com. 24 October 2022. www.space.com/james-webb-space-telescope-pillars-creation-excites-astronomers

3 Dennis Hung. The Evolution of Medical Imaging and Its Impact on the Healthcare System. www.healthworkscollective.com/evolution-medical-imaging-impact-healthcare-system/

4 Yuki Takechi-Haraya, Yukihiro Goda, and Kumiko Sakai-Kato. Imaging and size measurement of nanoparticles in aqueous medium by use of atomic force microscopy. Anal Bioanal Chem. 2018 Feb;410(5):1525–1531. doi: 10.1007/s00216-017-0799-3. Epub 18 December 2017.

5 International Maritime Organization. Resolution A.1149(32) Adopted on 15 December 2021 (Agenda item 9(a)) Revised Strategic Plan for the Organization for the Six-Year Period 2018 to 2023. A 32/Res.1149 28 January 2022.

6 David Sumi. V and E Bands Key to the 5G Revolution. July 2019. BroadbandCommunities Magazine, www.bbcmag.com/rural-broadband/v-and-e-bands-key-to-the-5g-revolution

7 mm wave frequency bands. RF Wireless World. www.rfwireless-world.com/Terminology/mm-wave-frequency-bands.html

Chapter 2

Engineering Sensors

Engineering sensors have evolved to support multiple roles and functions across the entire ship, with many active and passive sensors designed to measure vessel systems and critical infrastructure parameters and detect abnormal conditions. This is distinguished from sensor systems used on the bridge to assist with tasks such as navigation, safe passage and communication with other vessels and those designed primarily for monitoring passenger and crew activity, and cargo conditions. However, there is an overlap in rudder position, propeller shaft speed, thrust indication and other sensor applications where both engineering and deck seafarers have shared interests.

This chapter considers the various tasks performed by sensor systems having to do primarily with traditional engineering functions onboard ships. Different types of sensors typically found on most ships are described along with performance characteristics that help to determine whether the machinery, other apparatus, or processes they monitor are operating within or outside of nominal limits, or detecting events as they occur. The topic of "smart sensors" is discussed highlighting their use and advantages towards increasing safety and improving performance. Finally, shipboard process automation is discussed, emphasizing sensor fusion and how these systems can extend and enhance sensor performance.

2.1 Ship Function, Health and Performance

Engineering systems cover many functions with unique modalities and characteristics. These are generally associated with engine, propulsion, steering, electrical, heating, ventilation and air conditioning (HVAC); distillers, condensers, evaporators and associated pumps, valves, and piping. Control and health monitoring of engineering systems distributed throughout the vessel is accomplished through the use of a network of actuators and sensors to monitor nominal machinery performance and generate alarms resulting from out-of-tolerance and abnormal conditions. Vast improvements have been made in monitoring engines and drive trains, bearings, and other components that consider temperature, wear and vibration, water infiltration into

DOI: 10.1201/9781003377900-2

oil as well as monitoring exhaust gas and cylinder liners. Significant advances have also been made using new technologies for level sensing, water ingress, pressure monitoring, winch torque and load monitoring, and many other engineering sensing applications.

In addition to the above, other inputs from the engineering environment include motion, moisture, torque, velocity, acceleration, pressure, audio, and video. These inputs are often represented as a function of range, intensity, volume, resistance, conductance, and other characteristics of mechanical and/or electrical (analog and/or digital) signals created by sensors and sent directly to displays or indirectly using one or more data communication buses. Such functions form the underpinnings of engineering sensors for onboard systems that indicate system performance characteristics such as vacuum, quantity, liquid and gas flow, voltage and electrical current flow. The wide range of measured parameters and characteristics of data obtained from sensors used in engineering applications prohibit detailed discussion short of a full text dedicated exclusively to this task. However, generalizations can be made based on sensor function.

2.2 Types, Functions and Applications

Sensors are recognized primarily according to the things they measure. A partial list includes:

- Pressure
- Temperature
- Humidity
- Position
- Rotation
- Speed
- Velocity
- Distance
- Acceleration
- Angle
- Direction
- Force
- Strain
- Torque
- Vibration
- Voltage
- Current
- Resistance
- Capacitance
- Impedance
- Power
- Frequency
- Count
- Fluid level
- Fluid flow
- Fluid capacity
- Fluid remaining
- Fluid properties
- Heave/Pitch/Roll
- Air flow
- Audio
- Video
- Motion
- Noxious gas
- Time
- Date
- Open/Closed
- On/Off
- Error
- Other

Some sensors have the capability to measure multiple characteristics. Table 2.1 provides an overview and summary of various engineering

Table 2.1 Example Engineering Sensor Functions[a]

Function	*Signal Descriptions*
Alternating Current	Voltage, Current, Power, Frequency, Average Frequency, Average L-N Voltage, Average L-L Voltage, Phase A Frequency, Phase A L-N Voltage, Phase AB L-L Voltage, Phase B Frequency, Phase B L-N Voltage, Phase BC L-L Voltage, Phase C Frequency, Phase C L-N Voltage, Phase CA L-L Voltage.
Direct Current	Current, Voltage, Battery Temperature, Battery State of Charge, Battery Time Remaining, Ripple Voltage, Power.
Electrical	Resistance, Conductance, Impedance.
Electric Distribution	Switch/Breaker, Breaker Current, Hardware OFF Counter, Hardware ON Counter, Hardware ERROR Counter, Hardware ON Timer, Hardware OFF Timer, Hardware ERROR Timer.
Engine Monitor	Order, Response, Boost Pressure, Rotations per Minute (RPM), Hours, Coolant Pressure, Coolant Temperature, Exhaust Gas Temperature, Fuel Pressure, Fuel Rate, Oil Pressure, Oil Temperature, Tachometer, Tilt/Trim, Voltage, Fuel Economy, Fuel Consumption, Percent Load, Percent Torque, Trip Fuel Used, Vibration, Warning Indicator.
Fluid Flow	Flow Rate, Trip Volume.
Fuel Management	Distance to Empty, Time to Empty, Total Fuel Capacity, Total Fuel Level, Total Fuel Consumption, Total Fuel Used, Total Fuel Economy, Total Fuel Rate, Total Fuel Remaining.
Hull	Opening Status, Acceleration, Stress, Heave, Roll, Pitch.
Humidity	Inside, Outside.
Indicator	Status, Hardware OFF Counter, Hardware ON Counter, Hardware ERROR Counter, Hardware ON Timer, Hardware OFF Timer, Hardware ERROR Timer.
Mechanical	Force, Strain, Torque.
Motion	Acceleration, Angle, Angular Velocity, Angular Acceleration, Distance, Rotational Rate, Velocity.
Pressure/Vacuum	Barometric Altimeter, Compressed Air, Engine Boost, Engine Coolant, Engine Fuel, Engine Oil, Hydraulic Oil, Steam, Transmission Oil, User Defined, Water.
Propeller	Rotations Per Minute (RPM), Pitch, Order, Response.
Rudder	Angle, Order. Response.
Sound	Level, Decibels, Direction.
Tanks	Capacity, Level, Remaining, Properties.
Temperature	Battery, Engine Coolant, Engine Oil, Engine Room, Exhaust Gas, Freezer, Heat Index, Heating System, Inside, Outside, Refrigeration, Sea, Transmission Oil, User Defined, Wind Chill.
Thrusters	Status, Direction, Thrust %/RPM.
Time	Local Date, UTC Date, Moon Phase, Sunrise, Sunset, Local Time, UTC Time, Twilight AM, Twilight PM.
Transmission	Gear, Oil Pressure, Oil Temperature, Warning Indicators.
Vessel	Heave, Pitch, Roll, Trim Tabs, Stabilizers, Keel.
Video	Visible Imagery, Infrared Imagery.

Note:

[a] Derived in part from National Marine Electronics Association (NMEA) 2000/IEC 61162-3. Maritime navigation and radiocommunication equipment and systems – Digital interfaces – Part 3: Serial data instrument network.

applications and the characteristics measured in monitoring specific functions. Some overlap exists between functions for the sake of clarity and comprehensiveness.

Although there is great commonality among vessels, this table should be viewed as representative and general in nature due to the wide variety of ships and unmanned vessels and vehicles, their diversity in form and function and the extensive range of tasks and missions they perform. Engineering-related applications in which sensors are used can be classified into three general categories: ship hull and superstructure; engines, propulsion, and generators; and electrical, HVAC, and other auxiliary systems. Some of the locations on the ship where engineering sensor use is common include the hull (strain, fatigue, damage, motion, acceleration, opening status), service and cargo tanks (level, pressure, capacity), cargo holds (level, gases), engine room (temperature, pressure, vacuum, speed, flow, position, sound, video, consumption, load, torque, time, gases) and enclosed spaces (carbon monoxide, oxygen, explosive gases).

2.3 Performance Parameters

While a sensor detects and responds to physical input from the environment to produce an output (generally electrical), measuring a physical input and converting it into data can bring loss or distortion that can change the value or meaning of what is measured. These parameters can be viewed as static or dynamic; static is generally associated with a steady state input and dynamic with a changing state input.[1,2] Common parameters include:

- *Transfer Function*: A mathematical function that models sensor output for each possible input.
- *Linearity*: The deviation of sensor output from a straight line based upon a straight-line input.
- *Range*: Span of inputs the sensor can measure with a specified accuracy.
- *Sensitivity*: Ratio between change in sensor output to change in sensor input.
- *Hysteresis*: The degree to which a sensor output lags behind changes in the sensor input.
- *Repeatability*: The ability to consistently obtain the same output when applying the same input.
- *Drift* (also *stability*): Ability to maintain the same output when the input does not change over time.
- *Accuracy*: The degree to which the result of a measurement conforms to the correct value.
- *Resolution*: Smallest possible input that is detectable by a sensor.
- *Threshold*: Amount of change required at an input before a sensor changes the output.

Static and dynamic parameters are often a reflection of the quality of the sensor. Sensors can also be classified as being either analog or digital. Analog sensors produce an output that is continuous over time in proportion to the input, whereas digital sensors produce pulses or digital words corresponding to the input. Analog and digital sensors, where the output alternates between two fixed levels, are binary sensors that effectively function as on-off switches. Analog signals are usually converted to digital for transmission and analysis.

2.4 Engineering Sensors

Ships come in all sizes, exist for different purposes and perform a wide variety of functions. It is only natural that ship sensors follow the same pattern. There are many different sensors; some perform the same functions yet look completely different, while others look alike but perform different functions. The following paragraphs discuss the most common sensors used onboard ships of all types.

2.4.1 *Fluid and Bulk Level*

Sensors measure the levels of different fluids in tanks, bulk cargo in holds and elsewhere onboard ships and are designed according to the characteristics of what they measure. These include potable, ballast, bilge and black water; hydraulic fluid, fuel, lubricant oil, waste oil and liquified gases such as ammonia and natural gas. Some of the different types of fluid level sensors are listed below:

- *Resistive*: One of the more basic sensors is simply a resistive element attached to a float, lever, or other device where the value of electrical resistance changes as the device moves up or down. Some types of fluid level sensors may be stacked together with several in line for deep tanks beyond the effective sensing range of a single sensor.
- *Hydrostatic*: These sensors measure liquid column static pressure that varies in proportion to weight to calculate fluid level.
- *Capacitance*: Radiofrequency (RF) signals vary due to dielectric changes with the rise and fall of fluid level using a transmitter mounted externally to the tank.
- *Radar*: Microwaves are directed downward from the top of a tank or vessel, reflected back from the surface and then measured form the basic principles involved in Radar-based level sensors.
- *Ultrasound*: Sound waves are emitted, and the reflected waves from the surface are measured, with the transit time being proportional to the distance traveled.

In addition to cost, the key considerations for determining the proper type of level sensor to be used in any application include the characteristics of the material to be measured in terms of solid, liquid, slurry, or powder; the pressures and temperatures at which the material is kept in the container or vessel, flammability and explosion potential, and the nature and characteristics of the container or vessel itself within which the material is kept onboard the ship.

2.4.2 Fluid and Gas Flow

Flow can be measured as the mass or volume of a fluid or gas that passes through a particular point in a pipe, duct, or other conduit or structure. This is generally accomplished by taking a velocity measurement combined with a factor that considers the cross-section at the area where the measurement takes place. Additional factors that must be considered are the density and viscosity of the fluid or gas, which can vary widely and contribute to variations in flow measurements. Density is defined simply as mass per unit volume. It is thought of as the "weight" being measured based on the number of molecules occupying a given space (volume), whereas viscosity is the internal friction or resistance that exists within the internal layers of a fluid or gas. Both density and viscosity decrease as the temperature increases. However, density follows a fairly linear relationship to temperature, while viscosity mostly has an exponential relationship with temperature. Flow rates are expressed in units of volume or mass.

Fluid level and flow measurements are complementary and help determine how much of something exists and is available to go over a distance or for a period of time. Flow measurements are used in determining fuel oil consumption, monitoring engine coolants, controlling draught, trim and list through ballast water movements; balancing water systems, monitoring seawater inlets and monitoring the onload and offload rate of gas and liquid cargo transfer.

Different types of gas and fluid flow sensors most commonly found on ships include:

- *Thermal Anemometers*: This sensor operates based on the relationship between the amount of heat removed from a heated temperature sensor and the velocity of the gas or fluid passing over the sensor.
- *Differential Pressure*: Used for both gases and liquids, flow rate measurements are based upon the pressure drop across measured points, which is proportional to the square of the flow rate.
- *Ultrasound*: Sound waves are used to determine flow using either of two methods. One method is based on frequency shifts (Doppler effect) caused by liquid flow, whereas the other is based on the time it takes for sound to travel from one point to another based upon the direction and rate of the fluid being measured.

- *Laser Doppler*: This sensor uses a continuous wave laser combined with a laser detector to determine the change in wavelength caused by the fluid or gas being measured as a velocity function.
- *Mass*: The flow rate of gas or fluid mass is often measured to determine fuel-to-air ratios for proper combustion.
- *Vortex*: Flow measurements are based upon the creation rate of eddies and vortexes in proportion to the velocity of gas or fluid flow passing around a fixed object.
- *Turbine*: The gas or liquid flow rate through a propeller or turbine will cause it to turn or spin at a speed directly proportional to the flow.

Key considerations for determining the proper type of flow sensor include the characteristics of the gas or liquid to be measured in terms of purity, viscosity, density, liquidity or slurry, the pressures and temperatures at which they operate, and the properties and characteristics of the path in which the sensor is placed and operates.

2.4.3 Pressure

Pressure is measured by converting physical energy exerted against a surface into an electrical signal that is proportional to the force. Various types of pressure sensors exist using different technologies to make measurements. Pressure measurements are taken at various locations on a ship in support of functions such as engine fuel and lubricating oil, manifold vacuum and coolant; potable water, firefighting equipment, hull strain and fatigue,[3,4] shaft torque, and winch loads. Sensors commonly found on ships that are used to measure pressure include:

- *Piezoresistive*: Solid-state devices embedded within a sensor change their electrical resistance values in accordance with the pressure exerted on the sensor.
- *Piezoelectric*: Crystalline and ceramic devices embedded within a sensor that change their electrical polarity in accordance with pressure exerted on the sensor.
- *Variable Capacitance*: Materials embedded within a sensor vary in their ability to store an electrical charge according to pressure exerted on the sensor.
- *Variable* Reluctance: Materials embedded within a sensor that vary in magnetic reluctance (resistance to magnetic flux) in accordance with pressure exerted on the sensor.
- *Fiber Optic*: Fibers embedded within a sensor that change their optical characteristics to enable measurement of the phase differences between light signals that have taken different paths.

The four kinds of pressures include gauge, differential, absolute, and vacuum. All are measured in units of pounds per square inch (psi) or bars. Gauge pressure is measured in relation to atmospheric pressure. It is positive (above zero) when the pressure is greater than the ambient pressure (14.7 psi/ 1.01 bar at sea level) and negative when below ambient pressure. Differential pressure is a relative measurement of the difference in pressure between two measured points. Absolute pressure is a measurement against an absolute vacuum. Vacuum pressure is a measurement of pressure that is below normal atmospheric (ambient) pressure.

2.4.4 Temperature

Temperature is an expression of how hot or cold something is compared to a scale. Temperature measurements are made throughout a ship include engine coolant, cargo, exhaust, passenger and crew quarters, and food freezers. Both contact and noncontact measurements may be taken. Examples of different types of temperature sensors include:

- *Electro-mechanical*: Two dissimilar metals are physically bonded together, forming one device. As temperature changes, the two metals different rates of expansion and contraction cause torsion and physical movement that can be calibrated to temperature.
- *Thermocouple*: Temperature measurements are made using two different metal alloys that produce a small electrical signal when the temperature changes at one junction with respect to the other.
- *Resistance*: Resistive sensors are composed of two or more bonded metals or conducive polymers that change in electrical resistance as the temperature changes. Resistance may increase or decrease with temperature, depending on the materials used.
- *Infrared*: Noncontact measurements of liquid or material temperatures may be performed using infrared sensors that sense heat (or energy) in direct proportion to temperature.
- *Capillary*: Liquid expansion or contraction properties within a reservoir are often used to complete an electrical circuit between two contacts.

Temperature is generally expressed in degrees Fahrenheit or Celsius. Water freezes at 32°F/0°C and boils at 212°F/100°C. The Rankine and Kelvin scales are based on the Fahrenheit and Celsius scales, referencing absolute zero (–459.67°F/–273.15°C).

2.4.5 Humidity

Humidity represents the concentration of water vapor that exists within a gas. Absolute humidity refers to the actual amount of water present in terms of weight or volume. Relative humidity is the ratio between the absolute humidity and the saturation point at the same temperature. Measurements of humidity on ships are performed to monitor condensation levels in sensitive cargo and spaces; heating, ventilation and air conditioning system performance; maintain environmental conditions for proper operation of sensitive equipment, determining conditions conducive to fire and combustion, and for weather reporting and forecasting. Common types of humidity sensors include:

- *Capacitance*: Humidity is measured based upon the effect of absorbed water vapor on the ability of a sensor to store an electrical charge. As water vapor increases, the capacitance also increases based upon changes in the dielectric constant (permittivity) of the sensor materials.
- *Resistance*: Similar to a capacitance sensor, humidity is measured based upon the effect of absorbed water vapor on the resistance property of a sensor. This is usually an inverse relationship to humidity, where resistance decreases as humidity increases.
- *Thermally Conductive*: Humidity is measured using thermistors to determine the difference in thermal conductance between ambient air and a dry air reference (e.g., nitrogen), resulting in a change in resistance directly proportional to humidity.

Absolute humidity is measured in units such as grams per cubic meter (g/m^3), whereas relative humidity is measured as a percentage compared to saturation.

2.4.6 *Position*

This term is defined within the context of engineering departments on ships as being the location of something in a particular space or relation to another object or reference. Position sensors are designed to detect movement and convert this into signals for transmission, reporting, or control. Examples include rudder position to port or starboard of the ship's centerline; throttle being open, closed, or somewhere in between; and the detection of limits such as high vs. low, on vs. off, up vs. down, open vs. closed, etc. They can be divided into linear, rotary, and angular categories. Several different types of position sensors are likely to be installed throughout most ships and vessels. Common types of position sensors include:

- *Synchro/Resolver*: A *synchro* converts an angular position into an electrical signal transmitted via wire connection to a remotely located *resolver*, where the angular and rotational position is physically reproduced.
- *Limit*: Limit sensors are generally electromechanical devices used to detect the presence or absence of an object. Their use is associated with limiting the travel of an object to within specified limits.

- *Resistive*: A resistive sensor can be either linear or angular, usually a potentiometer forming an adjustable voltage divider using a three-terminal resistor. The resistance value changes with the position of a brush on the resistor, resulting in a variable electrical output dependent upon position.
- *Proximity*: Several technologies can be used to detect the nearby presence of an object. A switch can be used for simple detection (near/not near), whereas photoelectric, inductive, or ultrasonic techniques can provide degrees of closeness or proximity information.
- *Magnetic*: Magnetic position sensors are noncontact sensors that detect a magnetic field and adjust an existing electrical output based upon the strength and direction of the field in proportion to the position.
- *Inductive*: An inductive position sensor interacts with a magnetic field to create an electrical output whose strength is proportional to the magnetic field.
- *Capacitive*: A change in capacitance value is detected to determine the position of a measured object.
- *Optical*: Optical position sensors transmit light from a source to a receiver, or transmit and receive reflected light from the monitored object. Information about object position is based upon changes in light characteristics, including intensity, wavelength, polarization, or phase.
- *Ultrasonic*: Similar to optical sensors, an ultrasound source is used instead of a light source.
- *Hall Effect*: This noncontact sensor detects variations in magnetic fields to determine orientation, especially on a rotating shaft or device. This sensor is commonly used for speed detection.

Position sensor units of measurement include angle (degrees), direction, speed (rpm), velocity, position, range, and limits.

2.4.7 Vibration

Monitoring machinery vibration is an essential element of detecting failures and predicting future component failures. Vibration often stems from misalignment, imbalance, and wear on engines, motors, couplings, pumps, and other machinery. Measurement is accomplished mostly through the use of strain gauges and piezoelectric accelerometers that can sense back-and-forth motion. Vibration can be considered as being of two types:

- *Axial*: Vibration that is in line with the thrust projected from the machinery.
- *Radial*: Vibration that radiates perpendicular to the line of thrust.

Units to express vibration measurement can take several forms.[5] When considering the repetitive back-and-forth motion (frequency), this can be expressed in cycles-per-second or Hertz (Hz). The amplitude between vibration peaks or displacement from a fixed position can generally be measured in meter (m), centimeter (cm), millimeter (mm), or micrometer (µm) units. The velocity achieved during the vibration cycle representing the changing rate of displacement can be expressed as distance per unit of time; or cm/S, mm/mS, or mm/µS. Acceleration may be defined as the changing rate of velocity per unit of time and can be expressed as m/S^2, cm/S^2, or mm/S^2. Sound levels resulting from vibration can generally be expressed in decibels (dB) or watts per square meter (W/m^2).

2.4.8 Gases

Exposure to hazardous gases and deficiencies in essential gases in confined spaces onboard ships has gained significant attention in the press and by regulatory authorities.[6,7,8] Hazard sources include leaking gases and liquids, cleaning fluids, combustion (e.g., welding), dust concentrations and decomposing organic matter. Among the most significant gases monitored are:

- *Oxygen* (O_2): Depletion of oxygen, especially in cargo holds, has been the cause of many deaths.
- *Carbon Monoxide* (CO): Caused by the accumulation of exhaust gases from combustion.
- *Hydrogen Sulfide* (H_2S): Usually caused by leakage of blackwater tanks.
- *Lower Explosive Limit* (LEL): Combustible gases such as methane and natural gas.

Many permanent installation and portable gas detectors are available for detecting these and other gas hazards onboard ships. Multiple gases are often detectable using the same device.

2.4.9 Fire

Fires on board ships are one of the most dreaded occurrences, costing many lives and billions of dollars in damage and losses annually. Many nations have adopted IMO regulations under the Safety of Life at Sea (SOLAS) convention that specifically address shipboard fire detection capabilities.[9] All ships are subject to fire, but car-carrying vessels, in particular, have experienced even greater exposure as a result of carrying vehicles that contain not only combustible fuel but also lithium-ion batteries that are subject to seemingly spontaneous combustion.[10] Three different types of sensors are available that are capable of detecting the characteristics of fire. These include:

- *Smoke*: Operation is based upon the detection of particulates caused by smoldering materials.

- *Flame*: Detects ultraviolet and infrared radiation emitted by flames.
- *Heat*: Measures temperature above a threshold where the presence of fire is likely to exist.

Smoke is the most likely killer of people as toxic fumes may be emitted over a period of hours before the presence of flame or heat can be detected. Smoke detectors use a photo-electric sensor to visually detect particulates or a radioactive isotope as an ionization source to measure electrical conductivity between particles. Flame detectors use optical sensors to rapidly detect ultraviolet and/or infrared emissions associated with fire. Heat detectors are temperature sensors calibrated to a threshold that corresponds to a probability of fire. They are particularly useful in environments laden with dust and where high moisture levels exist. Many fire detectors contain two or more combinations of sensors. In addition to spaces where fire detection equipment is mandatory, car-carrying vessels are also adding networks of fire detectors throughout their cargo decks to detect fires with greater specificity as to location to further speed response times and fire suppression efforts.

2.4.10 Microphones

Microphones installed in the engine room and in other locations that are usually unoccupied can provide great insight into adverse situations as they develop. Easily detected using a simple live microphone connected to a remote speaker are vibrations, bearing squeals, clanking noises and other audible clues to actual or pending equipment malfunctions with which engineering staff are already familiar but may otherwise go unnoticed in the absence of staff within these spaces.

Sound monitoring devices are also available that can continuously monitor and record vibration, sound and noise levels, and air overpressure. These devices may be equipped with extended dynamic range microphones that can detect sounds outside the human range of hearing, are water-resistant and ruggedized to withstand harsh conditions. Monitoring can be performed continuously for days and weeks at a time, and thousands of events may be recorded when equipped with sufficient memory. Some such devices are standalone units equipped with internal batteries, while others are operated under the ship's power with battery backup that is network-connected to supplement other alarms and system capabilities.

2.4.11 Video Cameras

The use of video cameras onboard ships is dramatically increasing due to security requirements and to help make up for reduced crew sizes in maintaining surveillance of deck areas and enclosed spaces. Much like microphones, video cameras can offer visual clues as to events when they occur.

Some cameras have extended bandwidth capabilities to cover both the visible and infrared spectrums, or two cameras with complementary capabilities can be installed. This can be advantageous for identifying machinery that is running hot. Video monitoring can be performed continuously and recorded for days or weeks at a time over many events, usually at a location remote from the cameras connected via a network.

2.5 Sensor Interfaces, Controls and Panels

All sensors must be connected to interfaces and data buses that distribute sensor data to their destinations to be processed and initiate actions appropriate to sensor measurements. These include controls, alarms, panels and other devices such as computers and controllers associated with all sensed phenomena such as point and continuous levels and flows, leak detection and overfills, hazardous and nonhazardous atmospheres associated with ship hull and superstructure, critical system infrastructure, passenger and crew monitoring and cargo operations.

At these endpoints and in between there exist system components critical to efficient and reliable operation of sensor networks. Beginning from the sensors themselves, these components include:

- *Cables and Connectivity*: The channels and conduits through which sensor data is communicated.
- *Power Supplies*: Delivery of electricity to provide power to sensors and sensor networks.
- *Signal Conditioning*: Amplification, filtering, level changing and digitization of sensor signals.
- *Displays and Computers*: Destination for sensor data processing, alarm generation and user interfaces.

Each of these topics is discussed in the paragraphs that follow.

2.5.1 *Cables and Connectivity*

Electrical and fiber optic signal paths in the form of wires, cables and other conduits form a critical part of ship infrastructure through which individual sensors communicate the signals and data they acquire. Some sensors are individually wired through interfaces or directly to receivers, signal conditioning and data processing elements; while others are connected to common data buses that share many sensors and exchange large volumes of many types of data distributed to converters, other processors and displays. Wireless and Internet of Things (IoT) sensors designed specifically for the widespread sharing of data are also becoming more common.

Several onboard networks comprise digital communication buses for data transfer between sensors and onboard computers. Internal ship communication is accomplished using one or more data buses that include local area network (LAN), RS232/RS485, NMEA-0183/IEC 11623, NMEA-2000/IEC 61162-3, controller area network (CAN), wide area network (WAN) and Bluetooth. These and some other standards coalesce under the IMO e-navigation initiative, where maritime services transition from conventional transmission methods to contemporary digital technologies by providing and exchanging relevant information and data in a harmonized and unified format.

A thorough conversation on this topic is warranted when discussing engineering sensors. However, since the scope of this topic also covers navigation, passenger and crew, and cargo sensors; this topic is covered separately in Chapter 6 under Sensor Data Communications.

2.5.2 Power Supplies

Provisioning of electrical power to sensors is generally accomplished using centralized communication centers located in one or more standard 19-inch racks within which power supplies reside connected via interfaces that provide signal conditioning and power to the sensors. Power inputs are obtained from the ship's power at 115 V/230 Vac with power outputs provided to sensors at 12–28 Vdc at current levels sufficient to meet requirements depending upon the number of sensors connected to the system. Voltage and current requirements vary according to individual sensor requirements. However, many sensor interfaces are standardized, greatly simplifying this problem.

2.5.3 Signal Conditioning

Electrical signals present at the sensors themselves generally need to be adapted to coincide with the signal levels and formats that are compatible with the receivers at the processors, computers, and displays that use and act upon sensor data. Sensor signals are often too weak to be transmitted across long distances, can also be noisy and must be filtered to be useable. Sensor data can originate from several different forms that must be reformatted into digital word formats.

An example of signal conditioning is in the case where sensors that are based upon resistance can operate across ranges from 100 ohms (Ω) to 500 kΩ or more, depending on the type of sensor used. These are very common sensors that measure temperature, strain, humidity, fluid flow, force, pressure, and acceleration. Resistive measurements are often converted into voltage levels by bridge circuits that use a pair of voltage dividers, with one containing known resistance values and the other containing the resistive sensor. The circuit output is taken as the difference in voltage potential between the two dividers. When the balance between the dividers is the same there is no voltage difference. However, when the resistance of the sensor changes, the output voltage changes as a function of the measured value. In most cases, the magnitude of the voltage is small and requires amplification. This analog voltage needs to be converted into digital format to be useful. Bridge circuits generally feature low power consumption in the microamp (µA) to milliamp (mA) range and are often integrated directly into the sensor.

Another example is capacitance sensors, where capacitance values are converted directly into digital words at the sensor using an integrated circuit (IC) with correction and compensation for offset, sensitivity and drift made according to the specific characteristics of the sensor itself. This applies to single capacitive sensors where both terminals are accessible and differential capacitive sensors. Capacitance sensors also feature low power consumption in the microamp (µA) to milliamp (mA) range.

In many cases, the signals generated by the sensor itself and available at its output terminals are not directly compatible with circuit path characteristics to transmit sensor data, nor are they at the proper levels or in the correct format. It is, therefore, necessary to further process sensor signals for them to be useful. Several more common types of sensor signal processing techniques are listed in the following paragraphs:[11]

- *Amplification*: When sensor signal strength is below a threshold where a receiver cannot directly acquire data, it is necessary to boost or amplify the signal power to a level and range that is compatible with receiver input requirements. This applies to analog signals where the sensor output signal range should match the full range of analog to digital (A/D) converter capabilities, and to digital signals to ensure appropriate logic levels are attained. This is also important to overcome signal loss where long transmission paths exist between the sensor and receiver, and to help increase the signal-to-noise ratio such that environmental noise is minimized. Impedance matching between in-line system components is necessary to minimize signal loss, and sufficient output drive current must be available to ensure the signal is usable. Thermocouples and strain gauges typically require amplification to be effective.
- *Attenuation*: The opposite of amplification, attenuation entails reducing or limiting sensor output signal strength for analog sensors that may saturate or overload receiver input capabilities. This specifically applies when the sensor output signal is stronger than the range required for A/D conversion.
- *Filtering*: Electronic filtering eliminates environmental and other noise from sensor measurement data. This can include 50/60/400 Hz signals from local power sources and other frequencies radiated from equipment nearby the sensor and along the signal transmission path. With respect to

the operating bandwidth of the receiver, low-pass filters reduce or eliminate signals at the low end of the frequency spectrum from local power sources, while high-pass filters eliminate signals that exist above the highest frequency that is necessary to be coded at a given sampling rate to fully reconstruct the signal. Low-pass and high-pass filters are used together to eliminate signals within a specific band of frequencies.

- *Isolation*: Electrical isolation of sensor signals may be required in cases where signal parameters are significantly outside the range of the receiver, when voltage or current spikes can be introduced by equipment switching and other circuits, or where stray AC or DC signals may leak into a sensor signal path that may cause damage to equipment. This can be accomplished through the use of capacitors, transformers, optoisolators, or optocouplers.
- *Excitation*: This involves using supplemental voltage or current sources to facilitate the proper operation of active sensors that may include resistance temperature detectors, bridge sensors and Hall elements.
- *Linearization*: Sensors that produce nonlinear signals with respect to the physical measurement often require a process to transform their output into a linear signal. This may be accomplished using either hardware or software.
- *Compensation*: Sensor compensation involves calibrating a signal with respect to offset and span correction over a specific measurement range. This may be accomplished using either hardware or software.

Other types of signal processing can be performed to ensure optimal performance and flexibility by manipulating a signal to prepare it for the next stage of processing. The result must be in compliance with display and computer interface requirements at the next level of integration for users of these data.

2.5.4 Displays and Computers

Once sensor signals have been generated and processed, they are routed to the location(s) where they are recorded, used to provide notifications and alarms; or undergo additional computer processing as needed to operate and maintain the vessel. Display of sensor data may be accomplished using a simple analog gauge or digital numeric display, or integrated into complex, multi-sensor displays to monitor a combination of sensors. Such displays are often placed in one or more standard 19-inch racks that include one or more power supply(s), level converters, and signal processors, central processing units, a multi-function keyboard, and a display unit providing measurement and alarm data. This may reside at a single point location, such as a control room, or be distributed throughout the ship based upon areas of responsibility.

2.6 Sensor Relationships and Engineering System Automation

With increasing ship digitalization and automation, the role of sensors is ever-growing as they form the heart of how onboard systems view their environment. In the engineering department alone, automation has been introduced to reduce costs, improve operations and reduce the occurrence of human error and therefore accidents. Areas served represent a wide variety of integrated control systems including monitoring alarms and implementing auxiliary control systems, power management, propulsion, ballast systems, cargo and heating, ventilation and air conditioning. Smart sensors form the core of what is making automation possible.

References

1 AllICData. What Are the Static and Dynamic Characteristics of the Sensor? allicdata.com. Updated 21 November 2021. www.allicdata.com/news/sensor/what-are-the-static-and-dynamic-characteristics-of-the-sensor.html

2 Tom Kenny. Sensor Technology Handbook. Chapter 1, Sensor Fundamentals. Elsevier, Oxford, UK. 2005. Edited by Wilson, Jon S., pp. 2–5.

3 Sudripto Khasnabis. Different Technologies to Measure Hull Stresses in Ships. Marine Insight, Naval Architecture. 16 September 2019. www.marineinsight.com/naval-architecture/different-technologies-to-measure-hull-stresses-in-ships/

4 Y. Takaoka, K. Nihei, P. Vargas, P. Aalberts, and M.L. Kaminski. Application of Fatigue Damage Sensors in the Monitas System. Offshore Technology Conference (OTC), Houston, TX, USA. 30 April 2010. https://doi.org/10.4043/20870-MS

5 ByJu's.com. Unit of Vibration. https://byjus.com/physics/unit-of-vibration/#vibration-units

6 Maritime Executive. Two Killed in Gas Leak Aboard Chinese Bulker. 23 April 2018. https://maritime-executive.com/article/two-killed-in-gas-leak-aboard-chinese-bulker

7 Maritime Executive. Police Inspect Sydney Harbor Cruise Boats After Passenger Fatality. 7 February 7, 2019. www.maritime-executive.com/article/police-inspect-sydney-harbor-cruise-boats-after-passenger-fatality

8 Cargo and Cargo Hold Ventilation. INTERCARGO, the Standard Club and DNV-GL. January 2021. www.dnv.com/maritime/publications/Cargo-and-Cargo-Hold-ventilation-guidance-download.html

9 IMO. Part IV. SOLAS Chapter II-2. Construction – Fire Protection, Fire Detection and Fire Extinction. International Maritime Organization. London.

10 Valdes-Dapena, Peter. Burned Ship Carrying Luxury Cars Has Now Sunk. CNN Business. 2 March 2022. www.cnn.com/2022/03/02/business/felicity-ace-car-ship-sunk/index.html

11 David Ashlock and Anjelica Warren. The Engineer's Guide to Signal Conditioning. National Instruments, 2015. https://download.ni.com/evaluation/signal_conditioning/20712_Benefits_of_Integrated_SC_WP_HL.pdf

Chapter 3

Navigation Sensors

Many sensors are dedicated to different aspects of the safe and efficient movement of ships through perilous waters and the avoidance of other vessels and hazards to navigation while en route to their destinations. Biological sensors provide human vision, hearing, touch, smell and other abilities that have formed the core ship navigation capabilities for millennia and continue to do so, with the human brain the ultimate destination for sensor fusion and interpretation. However, the inventions of mechanical and electronic sensors have vastly expanded the range, scope, resolution and accuracy of the human senses to better assist mariners in performing essential pilotage, dead reckoning, celestial and other navigation techniques to locate position, determine time, steer a proper and safe course and continuously monitor the voyage through completion. Examples of invention include the ancient astrolabe, back staff, chip log, leaded line and sextant right up to modern electronic sensors such as Radar, Sonar, Global Navigation Satellite Systems (GNSS), autopilots, and weather instruments so vital to planning and executing a voyage. This has also led to secondary uses of time-delayed sensor data found in nautical charts, Electronic Chart Display and Information Systems (ECDIS), Notices to Mariners and Coast Pilots.[1,2,3,4] The data these products contain have been compiled in the past from many diverse sensor technologies and multiple sensors of the same kind including single-beam and multi-beam Sonars, Lidar, aerial and satellite imagery.

Marine navigation has been described as a blend of both science and art, and that science can be taught but art must be learned from experience.[5] Ship sensors illustrate how science can improve situational awareness and greatly enhance navigation safety. The principles of what ship sensors can do and how they function have been the basis for many courses of study and practical demonstrations. It is also the focus of research in machine learning that attempts to replicate, improve and accelerate human methods for viewing and interpreting sensor data. However, the art of marine navigation stems from prudent discernment of sensor limitations considering the

DOI: 10.1201/9781003377900-3

types of vessels on which mariners serve and encounter along their voyages, and the environment through which they must pass. Such knowledge takes years to acquire and master. It is this art of navigation that the developers of deep-learning Artificial Intelligence (AI) technologies are attempting to master through the fusion of multiple sources of sensor data combined with vast resources portraying vessel and human behavior.

Sensor technologies continue to adapt to new navigation applications, and the evolution of advanced sensor capabilities promotes the invention of new and enhanced methods to navigate. Historical examples include the replacement of the leaded line with Sonar, bells with high-power air and electric horns, and speed logs and Long-Range Radio Navigation (LORAN) with GNSS. The use of electronic sensors in vessel navigation is expanding across all maritime frontiers, whether they advise crewmembers on conventionally staffed ships or guide and control all navigation functions on autonomous vehicles.

This chapter proceeds from the perspective of how ship navigation is achieved with the benefit of sensors rather than focusing solely on the sensors themselves, although details on sensors are indispensable to their proper understanding. A review of the different types of ship navigation establishes a historical basis for how these inventions have been and continue to be adopted. This includes a discussion of methods developed over the centuries to perform navigation, emphasizing the contributions of sensors toward extending the human senses to accomplish these tasks. Finally, an overview is presented of the many sensors commonly found on ships, along with a preview of future navigation sensors and sensor technologies looming on the horizon.

3.1 Sensor Use in Navigation

An integral part of the navigation process is the ability to avoid collision and allision by communicating the meaning behind actions taken in response to sensor indications from Radar and Automatic Radar Plotting Aid (ARPA), Sonar, Automatic Identification System (AIS), radio, audible and visual signaling to and from nearby vessels and shore-based operation and control centers. Underlying these principles is the need for precise knowledge of the vessel's relative position with respect to other vessels, landmarks and aids to navigation (AtoN), land masses and sea bed topography along the route to the intended destination.

Essential to any discussion about navigation is the requirement to maintain a proper lookout by bridge watchstanders based upon the International Maritime Organization (IMO) Convention on the International Regulations for Preventing Collisions at Sea (COLREGS) rule 5 that states:[6]

> Every vessel shall at all times maintain a proper look-out by sight and hearing as well as by all available means appropriate in the prevailing circumstances and conditions so as to make a full appraisal of the situation and of the risk of collision.

In practice, the lookout task has traditionally been performed by one or more watchstanders posted at all times on the bridge and, weather permitting, stationed forward in restricted visibility conditions. They are tasked with reporting on lights, vessels and small craft, landmarks, large floating objects and marine mammals, indications of shoal waters, hazards to navigation and fog signals likely to herald the risk of collision. Efforts to further vessel autonomy find increasing assistance being rendered to bridge watchstanders by independent electronic bridge aids to monitor sensor data, detect potential hazards and suggest possible courses of action to be taken. Such aids satisfy the need for sight and hearing and *all appropriate means* to fully assess the situation.

Information obtained from navigation sensors are interconnected using multiple workstations, providing overlapping and redundant capabilities through an Integrated Navigation System (INS). Such systems are a convenient means for seafarers to prioritize and display navigation information in an organized manner and check the validity, consistency, latency and integrity of the data. The products of these displays are used to assist in route planning and monitoring, collision avoidance, alarm monitoring and to enhance overall navigation situational awareness.

Since approximately 1995 navigation sensors have witnessed the development and introduction of new instruments and technologies as diverse as Digital Select Calling (DSC) in Very High Frequency (VHF) and High Frequency (HF) radio, Navigational Text Messages (NAVTEX), enhanced Long-Range Radio Navigation (eLORAN), Forward-looking Navigation Sonar (FLS), the GNSS, Global Maritime Distress Safety System (GMDSS), AIS and Satellite AIS (S-AIS), digital Radar/ARPA, multibeam Sonar, broadband satellite communications, ECDIS and the Internet.[7] The rate of improvement in existing sensors, the invention of new sensor technologies and their augmentation to achieve new functionality through the use of AI continues to increase at an accelerated pace.

3.2 Overview of Navigation Sensor Functionality

A comprehensive list of navigation-related sensors and sensor systems commonly found on commercial and special-purpose vessels, along with a summary of associated signal descriptions, is included in Table 3.1. However, standards do not exist to enable seafarers trained in the equipment of one manufacturer's product to be proficient in the same type of equipment produced by other manufacturers. The bridges of many naval, coast guard

Table 3.1 Example Navigation Sensor Functions

Function	*Signal Descriptions*
Automatic Identification System (AIS)	Maritime Mobile Service Identity (MMSI), Navigation Status, Rate of Turn, Position, Speed Over Ground, Course Over Ground, True Heading, True Bearing, UTC seconds, IMO Identification, Radio Call Sign, Vessel Name, Cargo Type, Ship Dimensions, Type of Positioning System, Location of Positioning System Antenna, Ship's Draught, Destination, ETA...
AutoPilot*	Heading, Direction. Data sources: GNSS. Rate Gyro/INS, Rudder Indicator, Speed Log...
Audio (Sound)	Decibels, Frequency, Direction.
Camera – Visible	Video, Range, Direction.
Camera – Infrared	Video, Range, Direction.
Compass – Gyro	True Heading, Bearing.
Compass – Magnetic	Magnetic Heading, Bearing.
Depth Sounder	Depth, Range, Video...
Electronic Chart Display Information System (ECDIS)*	Coastline, Own Ship's Safety Contour, Isolated Underwater Dangers, Isolated Dangers, Scale, Range, North Arrow, Units of Depth and Height, Display Mode, Display Base. Drying Line, Buoys, Beacons, other Aids to Navigation, Fixed Structures, Fairways, Channels, Visual and Radar Conspicuous Features, Prohibited and Restricted Areas, Chart Scale Boundaries, Cautionary Notes, Ship and Ferry routes, Archipelagic Sea Lanes, Spot Soundings, Submarine Cables and Pipelines, ENC Edition Date, Chart Update Number, Magnetic Variation, Graticule, Place Names.... Data sources: Electronic Navigation Chart (ENC) database, GNSS, AIS, Radar, ARPA...
Emergency Position Indicating Radio beacon (EPIRB)	Vessel ID, Vessel Type, Vessel size, GNSS coordinates, Contact Information ...
Forward-Looking Sonar	Depth, Range, Video...
Global Maritime Distress Safety System (GMDSS)	Call Sign, MMSI, Selective Call Number, Mobile Earth Station Identification Number (MESIN), Ship Earth Station Identify Number (SESIN), 406 MHZ EPIRB Unique Identification Number.

(Continued)

Table 3.1 (Continued)

Function	*Signal Descriptions*
Global Navigation Satellite System (GNSS)	Positioning, Navigation and Timing (PNT) data...
Heading Indicator	Heading.
HF SSB Radio	Audio, Imagery, Digital data, MMSI, GPS data...
Laser Radar (LIDAR)	Video, Range, Direction...
Microwave Radar	Video, Range, Direction...
NAVTEX	Navigational Warnings, Meteorological Warnings, Ice Reports, Search & Rescue, Pirate Warnings, Meteorological Forecasts, Pilot Service Messages, AIS Messages, SATNAV Messages, Other Electronic Navaid Messages and Navigational Warnings.
Automatic Radar Plotting Aid (ARPA)	Video, Range, Direction, Closest Point of Approach, Time to Closest Point of Approach, Alarms, Target Data...
Rudder	Angle, Order, Response.
Search and Rescue Transponder (SART)	X-band Radar Beacon
Satellite	Imagery, Audio, MMSI, GPS data...
Side-Scan Sonar	Depth, Range, Video...
Speed Log	Speed Through Water, Over Ground, Total Log, Trip Log.
VHF-DSC Radio	Audio, Imagery, MMSI, Digital data, GPS data...
Voyage Data Recorder (VDR)	Data from: GNSS, Speed Log, Gyro Compass, Radar, AIS, ECDIS, Bridge Audio, VHF, Echo Sounder, Main Alarms, Hull Opening Status, Hull Accelerations and Stresses, Rudder, Engine, Propeller, Thrusters, Anemometer, Weather Vane...
Weather	Temperature, Barometric Pressure, Humidity, Dew Point, Wind Direction, Wind Speed...

*Note: The instrument itself is not a sensor but uses sensor data to perform its navigation-related function.

and other military vessels also generally have the same navigation equipment as civilian vessels and often comply with the IMO Safety of Life at Sea Convention (SOLAS) requirements. It should be noted that military vessels are not necessarily required to operate in the same manner as civilian vessels in terms of compliance with COLREGS and SOLAS and often do not regularly utilize equipment such as GMDSS and AIS.

3.3 Types of Navigation

Different types of navigation are used depending upon the physical environment in which the ship is found: harbor, harbor approaches, inland, coastal and ocean environments. Common to all modern vessels is the use of satellite navigation, which in practice almost universal reliance is unwisely made for positioning information. All large ships and many smaller vessels also rely upon Radar navigation for relative positioning information and collision avoidance. Radio navigation has been transformed from its origins as a means of obtaining positioning information via Omega, LORAN, and direction-finding systems to become a key provider of inter-ship communication via AIS, in addition to retaining its original functions of exchanging intentions between ships' bridges to coordinate close passage, ship to shore communications and for emergency use.

Piloting for navigating in restricted waters based on relative positioning to geographic and hydrographic features continues to be of great importance. Dead reckoning based upon course, speed, current, leeway, steering error and other factors remains a vital component of position estimation. Celestial navigation using lines of position to the moon, planets and stars is seeing a resurgence in significance, especially in view of the many limitations and frailties encountered with satellite navigation technologies. With sensors playing increasingly vital roles across all types of navigation, overreliance upon one or even several of these technologies without proper correlation between their results can lead to reduced safety of navigation and even disastrous outcomes. The following paragraphs describe these types of navigation and highlight the roles sensors play in each.

3.3.1 Satellite Navigation

Most conversations on satellites for ship navigation focus exclusively on GNSS services. However, often taken for granted and overlooked is the indispensable role played by weather satellites in providing the information necessary for the planning, continuous monitoring and execution of voyages. The following paragraphs attempt to offer comprehensive insight into all uses of satellites to permit and facilitate safe navigation.

3.3.1.1 Global Navigation Satellite System (GNSS)

The GNSS is comprised of several constellations in Medium Earth Orbit (MEO) positioned between 3,000 and 12,500 miles and Geostationary Earth Orbit (GEO) positioned geostationary 22,236 miles above the Earth with each constellation containing between 20 and 32 or more satellites developed by different countries to provide positioning, navigation and timing

(PNT) services to government, commercial and private users worldwide.[8] They enable precise determination of location in three dimensions (latitude, longitude and altitude) at any given time, although maritime use is generally at sea level except for some inland navigation locations.

Satellite systems that form the GNSS have been developed and are operated by various countries across the world and include:

- Global Positioning System (GPS) – United States.
- Global Navigation Satellite System (GLONASS) – Russia.
- Galileo – European Union.
- BeiDou Navigation Satellite System – China.
- Indian Regional Navigation Satellite System (IRNSS) – India.
- Quasi-Zenith Satellite System (QZSS) – Japan.

GNSS operates primarily in the L-band (1–2 GHz) and S-band (2–4 GHz) frequencies. The method used in the operation of GNSS includes the transmission of signals from various satellites in each constellation, with each signal containing a message with precise satellite location and current time information. Position is determined through triangulation by measuring the time it takes for signals to travel from multiple satellites to the ship's onboard receiver. Modern GNSS receivers often receive signals independently from several different GNSS systems to improve reliability in the event problems exist with one system, and to enhance accuracy.

Continuous PNT information is provided directly to the ship's navigation, communications and other onboard systems through constant communication with satellites to enable tracking of vessel movements and determine velocity and direction. Projection of future positions based on the vessel's current trajectory helps to determine adherence to planned routing and to detect and avoid potential hazards to navigation that may include other vessels, shoals and insufficient depths, and potential conflicts with charted and sensed objects. GNSS also provides onboard VHF and High Frequency (HF) radio transmitters and satellite terminals with position data to be communicated in the event of an emergency to help speed rescue efforts, and many systems offer features to mark waypoints that assist in man-overboard recovery situations. Capabilities to provide messaging for emergency warning services are being considered for the Galileo, IRNSS, and QZSS constellations.[9] This includes features for danger notification and subsequent instruction to escape from disaster situations. Several commercially available communication devices also locate positions using GNSS and then connect to other satellites to send and receive messages.[10]

Many limitations exist in the implementation of GNSS technology that reduces its usefulness and allows it to be exploited for nefarious purposes by nation states and individuals. One of its main flaws is that it operates using relatively weak signals from space that can easily be overridden by spoofing that displays false position information to users and can lead vessels to stray into shoal waters or directly into the hands of pirates and adversaries. These weak signals can also be interfered with through jamming resulting in denial of service (DOS) attacks that render it useless. GNSS effectiveness can also be limited due to signal path obstruction between the satellite and antenna by ship structures, and multipath interference due to reflected signals from the many metal surfaces on ships that can lead to errors in position, speed, or velocity calculations. Accuracy can also be affected by sunspots and space weather events that influence atmospheric conditions, especially with the ionosphere. It can also be adversely affected by severe weather conditions such as heavy precipitation, dense cloud cover and sometimes even fog. Satellite geometry, where satellite configurations are altered as parts of the constellation go out of service, can also adversely affect accuracy.

An additional potential limitation of GNSS services has to do with the ability of national operators to selectively degrade or dither signals to reduce positioning accuracy to nonmilitary users. This was called "selective availability" for the United States GPS, where autonomous horizontal positioning accuracy was advertised to be no worse than 100 meters 95 percent of the time. This function was turned off in 2000 and officially discontinued in 2007 after the U.S. DoD demonstrated the ability to selectively deny GPS signals on a regional basis as needed in a military area of operations when U.S. national security is threatened.[11] Such capabilities are likely retained for other nation-state controllers of GNSS satellite constellations.

It should be noted that Differential GPS (DGPS) services can be used where further signal augmentation is necessary to achieve centimeter-level or submeter-level accuracy. With one receiver established at a precisely known location as a reference, another roaming receiver calculates its position based on satellite signals and compares this location to the known location.[12] The difference is applied to the roaming GPS receiver in real-time in the field using radio signals. Differential techniques may be used with other GNSS systems as well.

3.3.1.2 Weather Satellites

Increasing bandwidth for ship satellite communications allows many real-time weather services and long-term forecasts to be provided directly to vessels at sea. This data is increasingly important for efficient route planning, weather avoidance and better-informed voyage decision-making. Commonly reported conditions provided by the World Meteorological Organization

(WMO) and National Meteorological & Hydrological Services (NMHSs) worldwide include:

- Cloud cover
- Precipitation
- Temperature
- Wind speed and direction
- Sea surface temperature
- Wave height
- Atmospheric pressure
- Currents
- Buoy observations
- Sea ice concentration
- Earth observations

Satellite providers of Maritime Safety Information (MSI) in terms of meteorological forecasts and warnings and a variety of other services include Inmarsat/Viasat SafetyNET and Iridium SafetyCast, as well as relatively lower-cost solutions requiring small antennas provided by SkyMate, Sirius XM Marine Weather, and Baron WxWorx. (Trademarks and Service Marks are properties of their respective service providers.)

Weather satellites generally operate within the same frequency bands as communication satellites, including the L, S, C, X, Ku and Ka bands. Exceptions exist where some weather satellites use frequencies in the Extremely High Frequency (EHF) and millimeter-bands up to 100 GHz, and analog Automatic Picture Transmission (APT) signals transmitted at frequencies between 137 and 138 MHz.

Concern has been voiced on whether 5G global networks can interfere with weather satellites in terms of their ability to sense water vapor, temperature and other atmospheric phenomena vital to weather forecasting.[13] There is not yet a consensus on this issue, and the full effects of 5G will be determined in the future once the inevitable expansion of this technology becomes more prominent in the satellite realm.

3.3.2 *Radar Navigation*

Radar as a navigation tool is a specialized form of piloting that enables navigators to measure distances and bearings to chart objects, Radar AtoN, and other vessels. Radar provided game-changing technology in terms of new navigation capabilities for position identification and collision avoidance, especially under low visibility and night time conditions, qualifying this navigation technique as a unique and separate discipline complementary to other types of navigation. However, the complexities involved in its operation required extensive training to master its use. The dangers of inadequate training resulting in improper navigation were illustrated by a Radar-assisted collision in 1956 when the third mate had misread his screen and inadvertently steered *Stockholm* straight into *Andrea Doria.*

Over 1,600 lives were at risk, 51 lives were lost and *Andrea Doria* was sunk.[14]

Radar system designs have changed over the years making them more accurate with greater range and resolution, providing more information than in previous generations, and easier to use and interpret. Also being introduced are tools such as Doppler technology that instantly shows whether a target is approaching or moving away, and ARPA that can create tracks and calculate the target course, speed and closest point of approach. New derivatives of Radar technology are also being introduced in the form of Synthetic Aperture Radar (SAR), millimeter Radar (mmRadar), and Light Imaging Detection and Ranging (Lidar) that promise new levels of situational awareness for not only traditionally staffed ships but also unmanned and autonomous vehicles. The following paragraphs describe some of the key technical aspects of Radar-based sensors and their use in modern vessel navigation.

3.3.2.1 First Analog, Then Digital Radar

Early Radar systems produced high-power continuous wave signals that feature combinations of amplitude modulation (AM), frequency modulation (FM), and pulse-width modulation (PWM) characteristics to detect and locate targets. Distance is measured by determining the time required for the signal to travel from the ship's transmitter to a target and then return to the ship in terms of the amplitude of the reflected signal being proportional to distance with variation in frequency to distinguish between targets. Pulse width that varies with reflected signal strength is used to measure target speed.

The range and resolution of the signal are dependent in part on the transmission frequency, with lower frequencies generally resulting in higher range and lower resolution than higher frequency transmissions. Radar signals are highly directional, thereby enabling bearing measurements based on the direction from which they were transmitted. Signal pulse width helps to determine resolution, and the higher the frequency at which the pulses repeat, the better it can discern faster-moving targets. Phase shift and Doppler shift characteristics detected in the returned signal also help to determine target speed and direction.

Analog Radar systems were prevalent in early maritime applications and are generally being phased out in favor of modern digital technology that uses much less power at reduced radiation levels to achieve greater performance. The heart of analog and early digital Radar systems is a klystron vacuum tube used as a signal amplifier. The processing of returning analog signals reflected off-targets is accomplished by converting them into electronic signals compatible with filters and mixers, whereby various signal characteristics are examined to reveal target distance, bearing and speed. Results are displayed on an electronic screen using symbols and characteristics appropriate for the target.

Digital Radar technology can generate complex signals and transmit multiple pulse widths where the returned signals are sampled, converted to digital signals and analyzed using a wide array of digital signal processing techniques that provide far greater control and manipulation of data than analog techniques. This results in greater accuracy and resolution than analog systems and new insight not available using analog signal processing to better detect and classify moving targets and to distinguish them from sea clutter, atmospheric effects and stationary backgrounds. Modern digital Radar systems are fully solid state, eliminating the need for a klystron vacuum tube amplifier. They are also less susceptible to interference due to better available filtering techniques. Digital Radar technology can provide features not available using analog technology, including SAR that can deliver higher spatial target resolution than conventional digital Radar systems.[15,16] Additional functionality can also be provided whereby, with each sweep of the Radar antenna, dual progressive scan transmissions may be transmitted and processed to simultaneously display two separate Radar ranges that can be manipulated separately. Additional capability is also available to overlay information from other systems, such as AIS, onto the Radar display, further enhancing situational awareness.

3.3.2.2 Radar Operating Frequency Trade-offs

Many early marine Radar systems operated in the L-band (500 MHz–1.5 GHz) frequency range that provided a good mix of range and resolution considering the technological state of the art of the time. Longer wavelengths are less susceptible to attenuation due to precipitation, and wider beam widths cover a larger area than progressively shorter wavelengths found at higher frequencies. As technology progressed, successively higher transmission frequencies were adopted offering greater resolution at the cost of lesser range. However, the advent of digital Radars helped to extend range as digital signal processing techniques improved. L-band Radars provided the foundation for the development of Radar-assisted navigation and collision avoidance techniques currently in existence.

Shorter wavelengths and higher operating frequencies have established S-band (2–4 GHz) Radars as the workhorse of the maritime industry for decades. Increasingly lower costs combined with long range on the order of up to 100 miles, good resolution and a robust ability to penetrate precipitation and other weather events provide a reliable sensor system upon which regulatory advancements were established to enhance the safety of navigation worldwide. These are the first systems where automated tools for target tracking and positioning were introduced.

Today's state-of-the-art X-band (8–12 GHz) Radar systems operating at shorter wavelengths and higher frequencies than S-band and L-band systems provide excellent short-range capability. Combined with very high resolution and good weather penetration, their use is ideal for waterway and port close-in maneuvering and for search and rescue. They feature many advanced capabilities including target tracking, identification and future position prediction made possible through enhanced signal processing and sensor fusion techniques. The simultaneous use of the combination of S-band and X-band Radars on the bridge to provide complementary perspectives of short and long-distance vessel and landmark characteristics can achieve greater situational awareness than was previously possible. X-band Radar systems can display Search and Rescue Radio Transponder (SART) positions transmitted from vessels in distress.

The most advanced short-range Radar systems available today operate in the L-band (24–27 GHz) and have even higher resolution than S-band and X-band systems. They are ideal for use in densely populated marine environments, target identification, port and lock maneuvering and collision avoidance. Due to their short wavelengths inherent to L-band signals, small targets, buoys and other AtoN are easily detected and identified.

3.3.2.3 Radar-Like Devices

New inventions that include mmRadar and Lidar are being developed for use by Unmanned Surface Vehicles (USVs) to enhance precise close-in maneuvering capabilities. These technologies are likely to also be adopted for use in navigation by conventionally crewed vessels in the future as their utility and benefits are more fully developed.

Operating primarily in the W-band (75–111 GHz), mmRadar has been used for a variety of military purposes to provide terminal guidance for missile targeting as an integral part of aircraft and ground vehicle fire control systems and to assist in aiming automated guns in ship defenses against incoming missiles. While recognizing limitations associated with their short wavelength operating frequencies, there are many opportunities for mmRadar use on civilian ships and vehicles. These include collision avoidance with other vessels and allision avoidance with objects, especially in port navigation, for real-time tracking of small boats in close proximity, for human in-the-water search and rescue, and ice navigation. The ELVA-1 mmRadar system for shipboard use is currently being marketed as capable of eliminating close-in blind spots and having a useful range of between 0 and 600 m.[17]

Lidar utilizes lasers that operate in the higher range of infrared light frequencies up to 400 THz. Its uses mirror those of mmRadar except with higher resolution due to the shorter wavelengths involved. Experiments have shown that ship position and heading angle were estimated using Lidar with higher accuracy than GPS, QZSS and inertial navigation, with a position estimation of approximately 1.2 m.[18] The Ouster OS2 Lidar system operates at a frequency of 345 THz (865 nm) and detects 10% reflective targets at 200 m with a maximum range beyond 400 m.[19]

3.3.2.4 *Piloting by Radar*

Piloting by Radar is the act of navigating a vessel through restricted waters using electronic eyes that can see relative positions only from the perspective provided by the interpretation of radio signals reflected from the local environment. The same practices that apply to piloting by eyesight also apply to Radar piloting. Still, their implementation requires learned skills and thought processes demanding intimate knowledge of Radar capabilities and limitations. Of particular interest are Radar AtoN, which should be specifically noted in the voyage plan. Terrain features, landmarks and AtoN that are easily verified visually in good weather can be distorted or omitted entirely from a Radar screen. For example, at long distances, low-sand beaches and shoals that are easily detected by the eyes of a mariner may not be detected at all by Radar unless covered by vegetation.[20] Structures on low sand beaches may appear to be other ships. Shadows in Radar returns caused by land features can significantly distort shoreline configurations and hide landmarks prominent to the eyes. Indirect or false echos caused by reflections from vessel structures may also appear on the screen and cause confusion. Other phenomena that may be encountered include side-lobe effects that cause circles or a series of echos around targets, multiple trace echos received from targets beyond the Radar range setting, and interference from other ship Radar systems and electronic emission sources that are in close proximity.

3.3.2.5 *Radar Collision Avoidance*

Whereas a major premise of the previous paragraph is that the relative positions of objects present on the Radar screen are fixed, the relative positions of most moving targets are constantly changing and require extreme due diligence. The monitoring of safe passage and evaluation of the risk of collision is covered under COLREGS rule 7, Risk of Collision, which states in part:[21]

> proper use shall be made of Radar equipment if fitted and operational, including long-range scanning to obtain early warning of risk of collision and Radar plotting or equivalent systematic observation of detected objects.

This is a requirement that Radar shall be used under all visibility conditions, day and night. The rule continues with,

> assumptions shall not be made on the basis of scanty information, especially scanty Radar information.

This part of the rule specifically acknowledges the limitations of Radar that make it unable to provide a complete or even an adequate representation of events. All decision-making should consider Radar as only one source of information that must be verified through other means such as visual contact, AIS, and bridge-to-bridge exchange of intentions.

3.3.2.6 Radar Beacons

The use of a Radar beacon (Racon) to enhance the presence of physical AtoN is accomplished by adding a transmitter that, when triggered by a Radar signal, will transmit a coded reply consisting of a series of dots and dashes to the interrogating X-band Radar. This reply appears on the Radar display in a line emanating radially from just beyond the echo of the AtoN. Racons may be used on both laterally significant and non-laterally significant aids alike; the Racon signal itself is for identification purposes only and therefore carries no lateral significance. They may also be used as bridge marks to mark the best point of passage.

3.3.3 Piloting

The term "piloting" refers to the navigation of a vessel through restricted waters. This involves one or more properly credentialled crewmembers, and sometimes a pilot knowledgeable of local conditions, present on the bridge performing the required tasks to achieve a successful transit. Integral to this process is proficient knowledge of sensor operation, interpreting and applying sensor data, and the real-time ability to combine and compare clues obtained through visual monitoring of the physical environment with multiple sensor readings to achieve and maintain comprehensive situational awareness.

The vast majority of sensors available on the bridge can assist in piloting navigation. Radar as a significant contributor to effective piloting has already been discussed in the previous paragraph. Other major contributing sensor sources include time-delayed data in nautical charts and ECDIS, and real-time data sources, including different types of Sonars, compass heading and bearing indicators, two-way audio hailers and even weather instruments. Of course, the human eyes, ears and other senses combined with seafarer training, experience and judgment are indispensable to successful piloting navigation. Unmanned and autonomous vehicles have additional sensors, supplemented by powerful computers containing machine learning and deep-learning AI-based firmware and software that attempt to replicate seafarer presence on board. Despite industry assertions to the contrary, as of 2024, the practical realization of such technologies is still years away into the next decade for general use and is only suitable now and in the immediate future for demonstration projects.

3.3.3.1 Historical Sensor Data

Piloting begins before getting underway with a thorough review of historical sensor data found in nautical charts, ECDIS, Light Lists, Notices to Mariners, Coast Pilots, Sailing Directions, tide charts, and other documents (both physical and electronic) of the correct versions, scales and resolutions appropriate to planning the voyage. These data have been obtained from a wide variety of sensors at some time in the past, and may or may not reflect actual conditions that may exist during the voyage but are representative of what might be expected at various points along the route. These data sources indicate the types and positions of AtoN, landmarks and other notable terrain and landscape features, in addition to hydrographic information including depth soundings and contours. Source diagrams and zones of confidence consist of graphic representations of the extent of hydrographic surveys included within nautical charts and accompanying tables of related survey quality categories. This information helps mariners assess hydrographic survey data and the associated level of risk to navigate in a particular area based upon a combination of the survey date, position accuracy, depth accuracy and seafloor coverage to detect objects on the seafloor.[22]

3.3.3.2 Visual Features and Landmarks

Once underway, the human senses supplemented with binoculars and other physical and electronic aids such as infrared (IR) and low-light cameras remain the standard for visual piloting navigation of conventionally staffed ships. Of specific interest to help accomplish this task are lighthouses, water towers, piers, terrain and any prominent objects or features displayed on navigation charts and contained within ECDIS to be used as AtoN that seafarers may correlate with visual sightings by taking compass bearings while en route. This includes critical buoys and lights whose characteristics may be verified as corresponding to their Light List entries, especially at night. The positioning of all visual AtoN should be cross-checked using Radar and GNSS whenever possible.

In the case of unmanned and autonomous vehicles, visible, low-light and infrared cameras are combined with microphones to help achieve situational awareness by sight and hearing through 360 degrees. However, seeing and hearing alone is insufficient without proper knowledge of what is being seen and heard. Appropriately trained and credentialed mariners remotely controlling unmanned vessels are constrained by and must rely entirely upon the perspective provided by shipboard sensors regarding the field of view, image and audio resolution and fidelity, signal path latency and other factors for decision-making. For example, a mariner having a lifetime of experience viewing the world through two eyes must adjust to a world where six eyes are the norm, assuming six cameras with a 60-degree field of view evenly distributed through 360 degrees, where each camera approximates a mariner's normal field of view. Fewer cameras could be used, but not without some loss of directionality and resolution, along with a distorted view to which the mariner is unaccustomed.

3.3.4 Radio Navigation

The earliest forms of radio navigation included now obsolete systems involving radio beacons, Decca, Omega, and Loran-C technologies. GNSS eventually replaced these systems, providing global coverage with far greater positioning accuracy. However, should GNSS services be denied or unavailable, few backup options exist to provide replacement PNT capabilities. The radio components and infrastructure are no longer produced or in place for long-range radio navigation except for time signal broadcasts over HF radio.

There still exist requirements for radio use, especially VHF but HF as well, for navigation purposes. In addition, several new developments have occurred regarding AIS capabilities to broadcast and display AtoN on Radar and ECDIS displays. These and other subjects are covered in the paragraphs that follow.

3.3.4.1 Radio Direction Finding (RDF)

Direction-finding has traditionally been accomplished using low frequency (LF) radio signals between 200 and 415 kHz and middle frequency (MF) signals between 550 and 1,500 kHz from radio beacons transmitting amplitude modulation (AM) signals commonly used for broadcasting stations and services. Dedicated maritime and aeronautical beacons transmitting Morse code identification signals, as well as commercial broadcast stations situated along coastal areas, would be received by the ship using a directional antenna to identify the compass bearing to the transmitter, which was then be used to guide the ship toward the station's location. RDF bearings are subject to inaccuracies due to quadrantal error caused by the influence of the local environment, coastal refraction due to differences in conductivity of the land and water, polarization error due to ionospheric activity and reciprocal error that may result in 180° ambiguity in the reading.[23] Radio beacons were generally phased out of use by the year 2000.

Direction finding is still used to locate the source of radio transmissions on 121.5 MHz, marking the position of an activated Emergency Position-Indicating Radio beacon (EPIRB). This signal is in addition to the satellite signal. It is accomplished while close proximity to the beacon and, presumably, lifeboats and survivors in the water near where an accident occurred on the water.

3.3.4.2 *Hyperbolic Navigation*

The Omega, Decca, and Loran C radio navigation systems began in the 1940s and 1950s and were based on using hyperbolic navigation principles that measured phase differences between different signals to generate lines of position from which a navigator could determine the position on a chart. Each system used different frequencies, with Omega operating on the Very Low Frequency (VLF) portion of the radio spectrum at 10–14 kHz, Decca operating on Low Frequency (LF) at 70 kHz and Loran-C operating on LF and Medium Frequency (MF) between 1.9 and 10 MHz. Each band of frequencies has its own unique signal propagation characteristics and limitations, with lower frequencies generally better able to cover longer distances. Omega accuracy was about 1,000 m at ranges of up to 10,000 nautical miles, Decca accuracy was about 100 meters at ranges of up to 400 nautical miles, and Loran C had an accuracy of about 1 nautical mile at ranges of up to 1,000 nautical miles (1,800 km).[24]

3.3.4.3 *Enhanced Loran (eLoran)*

eLoran has been considered as a possible backup system for GNSS. eLoran operates like to DGPS in providing differential correction to GNSS in local areas. It is currently not widely used except for research and testing in the United States and the United Kingdom, and it is in active use in South Korea. Testing performed by the General Lighthouse Authorities of the United Kingdom and Ireland (GLA) with eLoran experienced accuracies of between 8 and 10 m (95%) to seven ports on the east coast of the United Kingdom.[25]

3.3.4.4 *Automatic Identification System (AIS)*

AIS is used to provide vessel identification and other information for use in collision avoidance and general maritime domain awareness. Data is communicated in real time between ships using VHF radio channel 87B (Simplex), also known as AIS channel 1 (161.975 MHz), and ship to shore for VTS using VHF channel 88B (Duplex) as AIS channel 2 (162.025 MHz).[26,27] Other channels may also be used depending on national and regional regulations. Class A AIS generates output power of 1, 12.5, or 25 watts and is primarily intended for use by larger vessels of greater than 300 gross tons (GT) and all passenger vessels. Class B AIS, capable of generating an output power of 2, 5, or 12 watts, is intended for smaller vessels and recreational craft use. AIS messages comprised of digital data packets encoded with vessel name, Maritime Mobile Service Identity (MMSI) number, position, speed, course, type of vessel, ports of origin and destination, and other navigational information are continuously broadcast every 2–15 seconds between messages depending on vessel speed. AIS messages can also be received by satellite-based AIS receivers, allowing for global coverage and monitoring of vessel movements.

3.3.4.5 AIS Aids to Navigation (AIS-AtoN)

AIS can be used to augment the presence of a physical AtoN, such as a buoy, by providing a corresponding electronic presence at the same location on Radar and ECDIS broadcasting both laterally (e.g., Port Hand Mark) and non-laterally significant marine safety information (e.g., environmental data, tidal information and navigation warnings).[28] This is accomplished by fitting an AIS transmitter broadcasting using VHF radio on AIS channels 1 and 2 directly to the physical AtoN itself, installing one remotely within the line of sight of the physical AtoN and referred to as a synthetic AtoN, or virtually with installation in a location within line of sight to where a physical AtoN should exist but cannot be placed due to inaccessibility, consistently rough weather conditions or other environmental factors and referred to as a virtual AtoN. AIS AtoN can broadcast autonomously and at fixed intervals, providing the name, position, dimensions, type, characteristics and status from or concerning an AtoN.

3.3.4.6 Mobile Aids to Navigation (MAtoN)

Another type of AIS VHF radio-based AtoN is the Mobile Aid to Navigation (MAtoN), which is defined as a non-fixed or un-moored Aid to Navigation (AtoN) and does not include a fixed or moored buoy that is adrift from station, temporarily or otherwise.[29] MAtoN may exist in physical or virtual form and can be used for Ocean Data Acquisition System (ODAS) (e.g., to gather data on currents and weather), wreckage (e.g., containers, debris), water quality and pollution monitoring equipment, dynamic guard zones and convoys, underwater operations, enhancing navigational safety during military operations (e.g., no sail zones during minesweeping, target, exercises areas), towed and deployed applications (e.g., cable laying), search & rescue applications and special events (e.g., swimming competitions). It is specifically noted that MAtoN should not be used for unmanned vessel applications.

3.3.5 Dead Reckoning

Dead reckoning is an ancient navigation technique for determining the present approximate position based on projecting past courses steered and speeds over the ground from a known past position.[30] Not considered in DR navigation are the effects of leeway in the form of lateral drift off course caused by wind, currents and other environmental factors for which course adjustments are made as appropriate. Error is cumulative in that the error inherent in the present position estimate builds upon that which existed in all previous position estimates since leaving a known accurate position. DR is useful for predicting the times of sunrise and sunset, arrival at landfalls, and the sighting of lights and prominent landmarks. This technique is still used in modern times to provide redundancy in navigation methods and cross-check the accuracy of celestial and electronic navigation sensors. However, the sensors involved and methods used to perform them have changed dramatically.

3.3.5.1 Traditional DR Instruments

Various devices are necessary to perform DR, including a properly calibrated magnetic compass, navigation chart, logline or chip log, accurate chronometer or timepiece, parallel ruler, navigation divider and a log book to record observations and events. The compass, corrected for magnetic deviation caused by the vessel's unique physical and magnetic characteristics, determines the ship's course relative to magnetic north. The navigation chart, in addition to containing information essential to navigation and magnetic variation in the angular difference between true north and magnetic north, is used to plot the ship's course, measure distances, and identify landmarks or reference points encountered and anticipated along the way. The logline and chip log are devices used to measure the speed of a vessel through the water based upon the amount of line played out from a spool over a period of time, or a device deployed in the water and timed as to how long it takes to move between two known distances such as the bow and stern of the vessel. A watch, chronometer, or other timepiece, or electronic time reference known to be accurate is essential to record the times of observations or events that have occurred throughout the voyage from which elapsed time and distance traveled may be calculated. A parallel ruler and navigation divider are used for plotting and measuring distances on the navigation chart.

3.3.5.2 DR using Inertial Navigation

Modern-day Inertial Navigation Systems (INS) use the principles of dead reckoning to estimate position and velocity using a combination of accelerometers and gyroscopes to measure the acceleration and rotation of a vessel along several axes, oftentimes nine or more. Calculating changes in vessel position and velocity from an initial known position and velocity is performed by integrating these measurements over time. It is updated at regular intervals to provide a new estimated position.

Similarly, as traditional DR with the absence of external references for known positioning and velocity in PNT data, INS suffers the same problem of progressively increasing errors in velocity compounded into even larger errors in position resulting from integration drift. These errors include accelerometer bias, gyroscope drift, sensor noise, scale-factors, g-sensitive-factors, nonlinearity sensitivity, packaging, gravity modeling errors, and environmental factors.[31] Unlike traditional DR, INS employs advanced signal processing, error modeling and filtering algorithms combined with sensor fusion with external references, including GNSS. However, inertial navigation errors accumulate over time due to sensor inaccuracies and external disturbances.

3.3.6 Celestial Navigation

The use of astronomy in terms of time, celestial coordinates and motion of the stars, planets, moon and the sun for ship navigation dates from before the ancient Egyptians and the ancient Greeks developed many of the celestial navigation techniques still in use today. The invention of the chronometer that made precise timekeeping possible became widely available in 1759, and the sextant that helped mariners precisely measure angles between astronomical objects and the horizon to determine latitude above and below the Equator originated around 1791.[32,33] Celestial navigation has been downgraded in importance due to the advent of modern GNSS and other technology. It is rarely studied except in an academic environment or used operationally except by a few sailors and hobbyists. Despite this trend, celestial navigation remains a viable means to provide an independent source of positioning information to confirm the accuracy of GNSS readings, especially considering its limitations in terms of degradation, DOS attacks, and spoofing events. It can also provide a viable position reference for Inertial Navigation System (INS) alignment.

3.3.7 Weather Instrument Use in Navigation

This chapter described satellite communications as a means to provide real-time weather services and long-term forecasts directly to vessels at sea and how this data is becoming increasingly important for efficient route planning, weather avoidance and better-informed voyage decision-making. In reality, seafarers have been acutely aware of the weather for centuries and how observations of its effects can benefit or hinder ship navigation. The barometer has long been used to measure atmospheric pressure and how changes and trends in this measurement can help predict changes in the weather. Accompanying the barometer is the thermometer that measures air temperature, which can also help predict weather changes; and the hygrometer and psychrometer that measures humidity and dew point, which can help predict the likelihood of precipitation and fog. The wind vane and anemometer to measure wind direction and speed can help to plan a safe route and predict the severity of storms and hurricanes. When properly calibrated and maintained, weather instruments can provide a seafarer knowledgeable in their use and limitations with unique and real-time insight into conditions directly affecting the safety of navigation that are likely to go unnoticed by general area forecasts, especially in environments where few networked weather sensors are available for professional weather forecasters to observe and base their weather forecasts.

3.3.8 *Other Sensor Data*

Tides and currents can significantly affect a voyage, and proper planning can use the effects of these phenomena to greatly enhance the safety and speed of a journey. This is accomplished by anticipating water depths along the route and avoiding shoal areas, avoiding hazardous currents and waves, especially during the transit of harbors, inlets and areas where whirlpools and other hazards may be prominent. Fuel efficiency can also be affected to take advantage of following currents and minimizing head currents to help achieve better fuel economy and realize cost savings and avoidance.

Predicted tide times and levels are available from many published tide tables, and historical, current information is often available from local or even national sources. Real-time tide and current data can also be available over radio links and the Internet to data buoys equipped with tide gauges to measure tides at the buoy location and current meters that measure the speed and direction of surface and subsurface currents. An example of current measurements is illustrated below from the Chesapeake Channel LBB 92 [cb1101] buoy located in Annapolis, Maryland, USA, measured at 0950 local time when the query was made:

Predictions (Depth: 22.0 ft). Data source: https://tidesandcurrents.noaa.gov/map/index.html?id=cb1 101

Time (LST/LDT)	*Max/Min*	*Speed (knots)*
12:18 PM	ebb	−0.52
3:12 AM	slack	–
7:06 AM	flood	1.06
10:54 AM	slack	–
2:42 PM	ebb	−1.02
6:36 PM	slack	–
8:18 PM	flood	0.25
10:06 PM	slack	–

Today's Currents
Speed: 0.34 kts
Direction: 35° (toward NE)
Currents measured at: 22.0 ft below the surface

Plotted data in both standard and metric units over a daily period of between 0000 and 2359 is also available as desired.

3.4 The IMO e-Navigation Initiative

e-Navigation is defined by the IMO as:

> the harmonized collection, integration, exchange, presentation and analysis of marine information on board and ashore by electronic means to enhance berth to berth navigation and related services for safety and security at sea and protection of the marine environment,[34]

and is intended to meet present and future user needs through harmonization of marine navigation systems and supporting shore services. The development and implementation of the concept is coordinated by the IMO as a global collaborative effort to enhance maritime safety and efficiency by using electronic navigation. The strategy for e-Navigation implementation involves the integration and use of information from various sources, including MF, HF and VHF radio, GMDSS, Inmarsat, AIS, DGPS, Long-Range Identification and Tracking (LRIT) and Enhanced Maritime Mobile Service (EMMS).[35]

Many technologies and services comprising e-Navigation have been discussed elsewhere in this chapter and in Chapter 7, Shore and Intership Communications.

References

1 Chip log. Bright Hub Engineering. Albany, NY, USA. www.brighthubengineering.com/seafaring/60582-what-is-a-chip-log/

2 Mariners Astrolabe. Mariners' Museum & Park. Newport News, VA, USA. https://exploration.marinersmuseum.org/object/astrolabe/

3 Backstaff. Mariners' Museum & Park. Newport News, VA, USA. https://exploration.marinersmuseum.org/object/back-staff/

4 Using Lead Lines to Collect Hydrographic Data. National Oceanographic and Atmospheric Administration (NOAA), USA. https://celebrating200years.noaa.gov/transformations/hydrography/side.html

5 Bowditch. The American Practical Navigator. Chapter 1, The Art and Science of Navigation. Pg. 1. Defense Mapping Agency Hydrographic/Topographic Center. Bethesda, MD, USA. Pub. No. 9. 1995 Edition.

6 International Rules for the Prevention of Collisions at Sea, 1972 (COLREGS) as Codified in the United States by Coast Guard Commandant Instruction M16672.2D Navigation Rules – International and Inland. Rule 5, Look-out. 25 March 1995.

7 R. Glenn Wright. Innovations in Electronic Communications and Navigation. Unmanned and Autonomous Ships. Routledge, Taylor & Francis Group. Table 1.1, p. 7. 2020. ISBN: 978-1-138-12488-6.

8 NIST GNSS. Time and Frequency from A to Z. National Institute of Standards and Technology (NIST), Physical Measurement Laboratory, Time and

Frequency Division. Gaithersburg, MD, USA. www.nist.gov/pml/time-and-frequency-division/popular-links/time-frequency-z/time-and-frequency-z-g
9 Keiko Shimazu et al. Emergency Warning Services via GNSS Signals. 2020 IEEE Aerospace Conference. Big Sky, MT, USA. Institute of Electrical and Electronics Engineers. March 2020. DOI: 10.1109/AERO47225.2020. 9172377.
10 Brent Rose. The Best GPS Communicator to Buy for Your Next Adventure. The Verge, VOX Media Network. 12 April 1019. www.theverge.com/2019/4/12/18306207/best-gps-communicator-hiking-trails-garmin-spot-somewear-bivy
11 Glen Gibbons, 2007. Selective Availability: Completely Dead. Inside GNSS. 20 September 2007. https://insidegnss.com/selective-availability-completely-dead/
12 ESRI 2023. Differential GPS Explained. ESRI. Redlands, CA, USA. www.esri.com/about/newsroom/arcuser/differential-gps-explained/
13 Alexandra Witze. Global 5G Wireless Networks Threaten Weather Forecasts. Nature. 26 April 2019. www.nature.com/articles/d41586-019-01305-4
14 Robert J. Meurn. *Watchstanding Guide for the Merchant Officer*. Third Edition. Cornel Maritime Press, 2014. Case Study: Andria Doria/Stockholm Collision. p. 177. ISBN: 978-0-87033-631-7.
15 C. Lindstrom, R. Christensen, J. Gunther, and S. Jenkins. (2022) GPS-denied navigation aided by synthetic aperture radar using the Range-Doppler Algorithm. NAVIGATION, 69(3). https://navi.ion.org/content/69/3/navi.533.abstract.
16 F. Ali, G. Bauer, and M. Vossiek, "A Rotating Synthetic Aperture Radar Imaging Concept for Robot Navigation," in IEEE Transactions on Microwave Theory and Techniques, 62(7), pp. 1545–1553, July 2014, DOI: 10.1109/TMTT.2014.2323013
17 Short-Range Marine Radar FAQ. ELVA-1 Microwave Handelsbolag. Stockholm, Sweden. 28 July 2022. https://elva-1.com/blog/e-band-short-range-marine-radar-faq
18 Sawada, R. and Hirata, K. (2023) Mapping and localization for autonomous ship using LiDAR SLAM on the sea. Journal of Marine Science Technology. https://doi.org/10.1007/s00773-023-00931-y
19 OS2. Ouster OS2 Long-Range High-Resolution Imaging Lidar. Ouster, Inc. San Francisco, CA, USA. https://data.ouster.io/downloads/datasheets/datasheet-rev7-v2p5-os2.pdf
20 Bowditch. The American Practical Navigator. Chapter 13, Radar Navigation, pp. 213–214. Defense Mapping Agency Hydrographic/Topographic Center. Bethesda, MD, USA. Pub. No. 9. 1995 Edition.
21 International Rules for the Prevention of Collisions at Sea, 1972 (COLREGS) as Codified in the United States by Coast Guard Commandant Instruction M16672.2D Navigation Rules – International and Inland. Rule 7, Risk of Collision, 25 March 1995.
22 NOAA 2016. How Accurate Are Nautical Charts? NOA Coast Survey. 8 April 2016. https://nauticalcharts.noaa.gov/updates/how-accurate-are-nautical-charts/
23 Bowditch. The American Practical Navigator. Chapter 13, Radio Direction Finding. p. 176. Defense Mapping Agency Hydrographic/Topographic Center. Bethesda, MD, USA. Pub. No. 9. 1995 Edition.

24 Google Bard. Response to Question Posed on 1 June 2023. https://bard.google.com/
25 Alan Grant and Dana Goward. 10 Answers about eLoran. GPS World. 11 April 2022. www.gpsworld.com/10-answers-about-eloran/
26 NATO, 2023. AIS (Automatic Identification System) overview. NATO Shipping Centre. MARCOM Northwood, UK. https://shipping.nato.int/nsc/operations/news/2021/ais-automatic-identification-system-overview
27 MarineTraffic, 2023. Everything You Wanted to Ask about AIS. MarineTraffic Blog., 2023. www.marinetraffic.com/blog/ais-faq/
28 United States Code of Federal Regulations, Title 33. Chapter 1, Subchapter C, Part 62.52 Automatic Identification System Aids to Navigation (AIS AtoN). www.ecfr.gov/current/title-33/chapter-I/subchapter-C/part-62
29 IALA Recommendation R1016 Mobile Marine Aids to Navigation (MAtoN). Edition 2.0, December 2020. urn:mrn:iala:pub:r1016. www.iala-aism.org/product/r1016/
30 Bowditch. The American Practical Navigator. Chapter 13, Dead Reckoning. pp. 113. Defense Mapping Agency Hydrographic/Topographic Center. Bethesda, MD, USA. Pub. No. 9. 1995 Edition.
31 Andrey Soloviev. Inertial Error Propagation: Understanding Inertial Behavior. Inside GNSS. 4 June 2022. https://insidegnss.com/inertial-error-propagation-understanding-inertial-behavior/
32 Robert T. Gould. (1921) The history of the chronometer. The Geographical Journal, 57(4), pp. 253–268.
33 Peter Ifland. Taking the Stars: Celestial Navigation from Argonauts to Astronauts. The Mariners' Museum. Newport News, VA, USA, 1998.
34 Strategy for the Development and Implementation of e-Navigation. International Maritime Organization. MSC 85/26/Add.1, annex 20. London, 1998.
35 e-Navigation Strategy Implementation Plan – Update 1. International Maritime Organization. MSC.1/Circ.1595. London, 25 May 2018.

Chapter 4

Cargo Sensors

Independent sensors unrelated to ship's systems dedicated to track cargos and ensure goods arrive at their destinations fresh and in good condition have become widespread throughout the industry. Much of this trend was initiated by cargo transport through containerization, which has helped simplify the cargo handling process and standardize shipment methods. Almost every product imaginable can be shipped using standard containers, including clothing, textiles, furniture, appliances, machinery, mechanical and electronic equipment, and construction materials. However, there are also other methods to transport cargo consisting of goods and products that cannot fit within the confines of a standard container, and many ships have been designed especially for this purpose.

Although the tracking of shipments has been one of the primary uses of sensors, additional sensor needs exist beyond mere tracking for perishable goods that require special handling and specialized containers to ensure temperature, humidity, lighting, vibration, and shock are maintained at acceptable levels throughout the journey. This includes fresh and frozen foods, fruits and vegetables, meats, dairy products, liquids, beverages, medicines and flowers. Even more unique and specialized sensors are needed for dangerous goods requiring unique handling and extraordinary precautions, such as flammable liquids, solids and gases; combustibles such as batteries; radioactive materials, oxidizing and toxic substances.

Sometimes, it is also necessary to instrument the cargo spaces and even the cargo itself to ensure its safety and that it is kept in good condition, to gain additional insight into unique and undocumented behavior properties, and to provide advance warning of possible risks to the ship in carrying such cargo. Examples of shipments that need specialized sensors to maintain safety and to help prevent the loss of the ship include car carriers transporting electric vehicles that use lithium-ion batteries subject to spontaneous combustion and dry bulk carriers whose cargo contents may be prone to liquefaction.

This chapter reviews some ship configurations that carry cargo and highlights those with the most pressing need for sensors independent of onboard engineering and navigation systems. Many of these sensors and their principles

DOI: 10.1201/9781003377900-4

of operation have been discussed in the previous chapters on engineering (e.g., temperature, humidity, vibration, accelerometers, etc.) and navigation (e.g., Global Navigation Satellite System – GNSS). Therefore, a major focus of this chapter is to consider the needs of cargos and ship configurations subject to the highest occurrence of accidents and casualties that threaten safety in terms of loss of life, compromise of ship stability, loss of the entire ship, environmental damage and subsequential economic impact.

4.1 Ships and Cargos

A brief glimpse is made at the different types of cargo ships that exist on the high seas and many smaller cargo ships involved in coastal and inland shipping. The most common types of cargo ships that account for the majority of the world's maritime trade are container ships, bulk carriers, tankers, Roll-on/Roll-off (Ro-Ro) vessels and special cargo or multipurpose cargo ships.

Container ships can be the largest of all ships, designed to carry thousands of standardized shipping containers that are easily loaded and unloaded. The largest of these are the sister ships *MSC Irina* and *MSC Loreto* that entered service in 2023, each with a length of 400 m (1,312 feet), a width of 61 m (200 feet) and a capacity of 24,346 Twenty-foot Equivalent Units (TEUs).[1] Evergreen's *Ever Alot* is also of comparable size at 396 m (1,300 feet) in length, 62 m (203 feet) in width and 24,004 TEU capacity. Many other ships with container capacities of thousands to less than one hundred TEU service smaller ports and communities.

Bulk carriers are ships designed to carry dry unpackaged cargoes featuring from one to several hi-volume cargo compartments and can also be listed amongst the largest of vessels. Their cargo includes large quantities of coal, grain, iron and other ores. One example is MS *Ore Brasil*, which measures 362 m (1,187 feet) in length, 65 m (213 feet) in width and its hull depth (not draft) is 30.4 m (115 feet).[2]

Tankers are also among the largest of vessels and come in various sizes, shapes and configurations designed to carry liquid cargos that generally include crude oil, refined oil products, gas and many types of chemicals. The largest of these is *Universal Leader*, a Very Large Crude Oil Carrier (VLCC) with a length of 336 m (1,102 feet) and a width of 60 m (197 feet).[3] Another tanker of a very different design is *Shinshu Maru* that carries liquid natural gas whose length is 399.9 m (1,309 feet) and width is 48 m (157 feet).

Ro-Ro carriers transport vehicles and other wheeled cargo that can be driven on and off the ship. These can range from small inland and coastal ferries that also carry passengers to large vessels transporting cars, trucks, railroad cars, cranes and other construction machinery to destinations worldwide. The largest of these is MV *Tonsberg*, with a length of 265 m (860 feet), width of 32 m (105 feet) and draft of 11 m (36 feet). This vessel has six decks and can carry heavy rolling and non-containerized cargo, breakbulk and cars.[4]

General cargo ships or multipurpose vessels are flexible in their use and can transport a wider variety of goods and cargos including containers, bulk cargo and Ro-Ro vehicles. There are many ships in this category, and the size leader by individual ship changes frequently. However, in 2019, the ten largest operators by deadweight of multipurpose-project-heavy-lift tonnage deployed a combined fleet of 476 ships with a total deadweight of 8,220,000 million tons and an aggregate lifting capability of 155,000 tons.[5]

Other vessels that transport specialized cargo include ships that transport other smaller ships, refrigerated ships for transporting fish cargo, livestock carriers, ships that transport large aircraft components such as fuselage sections, crane ships that operate at sea, open deck ships that transport giant cranes for port use, ships that recover space rockets after launch and transport them to port, and just about most anything else that can be moved from one place to another.

4.2 Containerized Cargo

In times past, the Captain has been held liable for what happens to the ship and environs due to being knowingly negligent in failing to take proper precautions to transport hazardous cargo and mishandling non-hazardous cargo. However, the situation is much less clear when operations involve a ship transporting thousands of containers at a time. In such cases it is virtually impossible for a ship's crew to inspect each container and they must rely on the accuracy of bills of lading, hazardous materials declarations and labeling to be aware of the onboard dangers of transported cargo. It is incumbent upon the shipper to guarantee the loading of individual containers is properly accomplished to safeguard their contents and that they are mutually compatible and not likely to come in contact, interact and combust. The problems associated with shipping containers are well documented, and the headlines are constantly filled with news stories about mishaps on land and at sea in which they are the source of risk and loss.[6,7,8] Some of the different sensors for the various uses that can exist onboard container ships are described in the following paragraphs.

4.2.1 *Container Hold Sensors*

Enclosed spaces within the hull designed to carry shipping containers are equipped with sensors to confirm the security of hatch covers. In addition to visual inspection by direct crewmember observation and video cameras to verify closure, other sensors may confirm or contradict the visual indications of whether a hatch cover is closed. One example is the use of pressure sensors to detect differences in air pressure between the outside and inside of the container hold, where the same pressure may indicate open and differences in pressure indicate closed. Magnetic sensors may also be distributed along the hatch and frame to detect gaps and determine if the hatch is open or ajar. Ultrasonic leak detectors may also detect changes in airflow or pressure that may represent air or gas leaks around the hatch covers.

Additional sensors are also required in container holds to detect and suppress fire.[9] These include temperature, smoke, and heat sensors to detect indications that a fire may be present. Also, sensors are required to monitor oxygen concentration within the space as part of a carbon dioxide (CO_2) fire suppression system to verify and maintain an inert atmosphere during and after a system discharge.

4.2.2 *Shipping Container Sensors*

In 1967, the International Standards Organization (ISO) and International Maritime Organization (IMO) began the process of container standardization that resulted in the adoption of the International Convention for Safe Containers (CSC) in 1972.[10] This convention requires shipping container doors to be marked with alphanumeric codes to assist in tracking shipments. The codes contain a unique identifier and owner information, the dimensions and type of container, operational marking representing maximum gross weight and container weight (tare), and a CSC plate verifying good condition and acceptable safety. This method made possible the identification and tracking of individual containers throughout their journey and even their location on the ship.

Initial efforts using electronic code readers on containers increased port efficiency somewhat but proved cumbersome by creating new needs to position these readers accurately to achieve desired results. In 2004–2005, efforts were undertaken to mark shipping containers with Radio Frequency Identification (RFID) tags to enable wireless tracking of shipping container movements and detect events such as door openings with reporting via local communications networks, mobile data services and satellites.[11,12]

In the following years, many factors have contributed to the development of "smart containers" comprised of a wide range of sensors to provide identification of the container and its contents and also to gather information regarding GNSS positioning, temperature, humidity, shock, and acceleration, the presence of explosive gases, smoke and many other factors resulting in vastly expanded data collection efforts. Sensors embedded within containers should also help to detect container fires onboard ships that have claimed many lives and resulted in the loss of ships and their entire cargos.[13,14] The occurrence of container ship fires are frequently reported in the news. A short list of examples include:[15]

- In 2018, the Ultralarge Container Ship (ULCS) *Maersk Honam* vessel caught fire, killing five crewmembers.
- In 2017, the 13,800 TEU *MSC Daniela* was on fire for more than a week.
- The 2016 fire on the 9,000 TEU *CCNI Arauco* in Hamburg, Germany.
- In 2012, a fire onboard the container ship *MSC Flaminia* burned for six weeks and killed three crew.

The containers collect and transmit data about their location, condition and cargo, which can be used to optimize and improve the efficiency of container routing. These data can also enhance security with sensors that can detect tampering, unauthorized access, theft of contents, and cargo damage due to shock; and promote sustainability of the shipping supply chain by tracking the progress of the shipment and to identify potential delays. This is especially needed due to the increasing number of containers lost at sea with resulting pollution and their widespread dispersal across the seas as large hazards to navigation. Smart shipping containers equipped with sensors, GNSS tracking and other Internet of Things (IoT) devices are now replacing the simple steel containers and discrete sensors of yesterday.

4.2.3 Parametric Rolling

A vessel behavioral phenomenon known as Parametric Rolling Movement (PRM) can cause ships to roll at extreme angles of up to 30–40 degrees or more with very little notice, even in low to moderate sea states. This is seen primarily with container ships and car carriers, both of which feature hull forms with flared fore and aft decks. Other types of vessels may be similarly affected in head or following seas. Container ships may experience exceptionally high container losses when combinations of rolling period, wave conditions, ship's speed and heading cause sudden and extremely rapid increases in roll motions, even in relatively mild wave heights.[16,17] Such conditions can be very hazardous and likely occur when wave lengths approximate vessel length and the rolling period is twice the wave encounter period.[16] This can result in extreme stresses on containers and cause failure of their securing system and loss of containers, and may possibly capsize the ship itself.[18]

The Danish company Kjærulf Pedersen reports it has developed a sensor system that can recognize the tendencies that lead to parametric rolling and send a message to the ship's controls so the system and crew have time to take corrective action. Based on its inclinometer sensor product designed to measure orientation using an accelerometer, gyroscope, and magnetometer; digital signal processing is used to perform inclinometer calculations.[19] Three sensors placed at the bow, amidship and stern record the ship's movements and compare observations with the ship's speed, acceleration and direction to detect whether the movements are approaching a state that can become uncontrollable.[20] Ship crews may then be alerted to promptly take measures such as course changes and speed adjustments to overcome this condition.

4.3 Bulk Cargo

The Intercargo 2019 Bulk Carrier Casualty Report provided an analysis of statistics covering 2010–2019, citing thirty-nine bulk carriers with 173 seafarers as having been reported lost over that period.[21] Cargo shift and/or liquefaction was identified as one of the greatest concerns and likely to be the main reason behind eight bulk carrier casualties and 106 seafarers losing their lives. Losses included six bulk carriers carrying nickel ore from Indonesia, one with iron ore fines (assumed to be laterite) loaded with high moisture content from Malaysia, and one with bauxite from Malaysia. These statistics were updated in 2023, citing that in the ten years since 2013, more than two-thirds of the 104 seafarers who perished in the 26 bulk carriers lost at sea died due to cargo liquefaction.[22] Other causes of bulk cargo ship losses included grounding as the most common with 17 losses, flooding for five ship losses, and four ships lost due to unknown causes accounting for 35 lives.

4.3.1 Cargo Liquefaction

A significant problem for bulk cargo ships is the fact that solid granular bulk cargo such as crushed nickel ore, iron ore fines, bauxite fines, mineral concentrates, and some by-products like "red mud" can behave like a liquid during transport sloshing around in the hold and leading to loss of stability.[23,24] A major vulnerability exists when the cargo undergoes liquefaction, moves to one side of the ship and solidifies, adversely affecting stability. Such conditions can occur when moisture content and degree of saturation of the cargo are sufficiently high, and a combination of conditions exist related to particle size distribution, pressure within the particle pore spaces, and compaction that can reduce friction between particles and allow them to flow and shift. One example attributed to cargo liquefaction is the January 2015 loss of *Bulk Jupiter* and 18 crewmembers carrying 46,400 tons of bauxite that sank off the coast of Vietnam, with only one survivor.[25] Unfortunately, there are also many other examples.

To help mitigate the risk, international regulations and guidelines such as the International Maritime Solid Bulk Cargoes (IMSBC) code provide standards and procedures for the safe transportation of solid bulk cargoes, including testing and moisture limits, certification of cargoes, monitoring moisture content and ensuring suitable ship stability calculations are performed before loading and during the voyage. Cargos known to liquify are listed as Class A under the code. Each type of solid cargo now has a Transportable Moisture Limit (TML) rating 10% beyond the product's Flow Moisture Point (FMP), the percentage of moisture in the product at which it may begin to behave like a liquid. Should the Moisture Content (MC) of the cargo exceed TML, it is not considered safe or suitable for shipment. Additional approaches to ensure safety during transport include installing sensors inside cargo holds to monitor moisture content, humidity and water pressure within the cargo. Lasers may also be used to monitor changes in cargo position to provide advance warning of unsafe conditions.[24]

However, the IMSBC code does not necessarily list all cargos that may liquefy. P&I Club Gard recently identified cargos such as fluorspar, certain grades of coal, pyrites, mill scale, sinter/pellet feed, etc., as also being able to liquefy and provide many references on the subject.[26] In their warning, they cite one incident where a vessel carrying over 1,900 tons of soil from a landfill listed eventually sank due to wind and waves, causing the cargo to behave like liquid during transit.[27] Another incident was a cargo of calcium carbonate that caused a vessel to develop a list shortly after leaving the load port. The cargo's moisture content was greater than 30%, exceeding the TML of 24% and Flow Moisture Point (FMP) of 26.7%, where the material consisted entirely of particles less than 2 mm.

4.3.2 Explosion

An explosion took place, and four crew members were injured on board the 120,600-dwt bulk carrier CSSC *Cape Town* as it entered British Gibraltar Territorial Waters in the Bay of Gibraltar on 19 February 2021.[28] The ship, built in 2020, was loaded with 112,365 metric tons of coal at the Curtis Bay Coal Terminal in Baltimore, Maryland, USA. According to the report, the explosion appeared to be in the area of the vessel's forecastle, and the cause of the explosion is not known. No fire was reported.

Coal is classed as a non-dangerous cargo under the Convention on Facilitation of International Maritime Traffic. However, coal dust suspended in the air can be highly explosive and very susceptible to spontaneous combustion because it has much more surface area per unit weight than lump coal. For an explosion to occur, five simultaneous elements must be present: fuel, heat, oxygen, suspension and confinement.[29] A fire may occur should fuel, heat, oxygen and confinement conditions exist in proper quantities. However, an explosion could occur with the suspension of burning fuel as may result from the introduction of a sudden blast of air.

The primary sensors for determining coal dust concentrations are cumbersome and involve frequent maintenance and replacement of filters. The dust can easily block and pollute photodiodes of light-scattering dust concentration optical sensors. New research in electrostatic induction coal dust concentration sensors has yielded high detection accuracy for coal dust and may be suitable for future use in bulk cargo ships.[30]

4.3.3 *Unpreparedness to Fight Fires*

Numerous opportunities can lead to shipping companies and crewmembers being ineffective in fighting ship fires. The July 2020 fire aboard the amphibious assault ship USS *Bonhomme Richard* (LHD 6) at Naval Base San Diego while it was moored pier side for maintenance is a good case study for such events. In a report released by the U.S. Navy describing the complete loss of the vessel, four categories of causal factors allowed for the accumulation of significant risk and led to an ineffective fire response.[31] These included the material condition of the ship, the training and readiness of the ship's crew, the integration between the ship and supporting shore-based firefighting organizations, and the oversight by commanders across multiple organizations. The command investigation also concluded that "a lack of familiarity with requirements and procedural noncompliance at multiple levels of command" contributed to the loss of the ship.

This incident did not involve a civilian bulk cargo ship, and there were many heroic acts of bravery, ingenuity and resourcefulness in the actions of sailors across the San Diego waterfront and others who had a role in the firefighting response. However, lessons learned from this naval equivalent must highlight the fact that gross incompetence and negligence at the highest levels resulted in most of the onboard smoke and fire detection sensor capabilities and 87% of the ship's fire stations remaining in an inactive equipment maintenance status that prevented the ship's aqueous film forming foam sprinkling system from being used to fight the fire.[32]

4.4 Liquid Tank Cargo

Cargo monitoring is an integral part of the engineering sensor suite for most ships built to specifications for specific purposes, such as oil tankers and Liquid Natural Gas (LNG) carriers. These sensors are generally fixed systems that aim to detect unsafe and explosive gas–air mixtures in cargo tanks and holds, ballast tanks and void spaces before they reach the Lower Explosion Limit (LEL). However, accidents with catastrophic results still occur. The following paragraphs discuss the unique aspects of some of these accidents to illustrate potential additional uses of sensors that may help to avert accidents in the future.

4.4.1 *Cargo Tank Overheating*

The UK Marine Accident Investigation Branch (MAIB) report on the investigation of the 28 September 2019 cargo tank explosion and fire on board the chemical tanker *Stolt Groenland* while alongside in the port of Ulsan, Republic of Korea, found that heated cargo tanks in nearby compartments likely caused dangerous warming of a temperature-sensitive cargo of styrene monomer.[33] Details of this investigation cited in the report include:

> The rupture of the styrene monomer tank resulted from a runaway polymerisation that was initiated by elevated temperatures caused by heat transfer from other chemical cargoes. The elevated temperatures caused the inhibitor, added to prevent the chemical's polymerisation during the voyage, to deplete more rapidly than expected. Although the styrene monomer had not been stowed directly adjacent to heated cargo, the potential for heat transfer through intermediate tanks was not fully appreciated or assessed. Critical temperature limits had been reached before the vessel berthed under the road bridge in Ulsan. The tanker's crew did not monitor the temperature of the styrene monomer during the voyage, and therefore were not aware of the increasingly dangerous situation.

This accident indicates a need for the use of one or more smart sensors capable of performing the task supposed to be accomplished by the crew to monitor the temperature of the styrene monomer and the environment in the immediate vicinity and to promptly analyze and report the findings to crewmembers. Several environmental characteristics, such as temperature, trace gases and other phenomena appropriate to styrene monomer, could be monitored simultaneously using multi-sensor elements within one instrument enclosure. In addition to monitoring specific environmental characteristics, integrating machine learning and deep-learning artificial intelligence (AI) technologies into the sensors can discern interrelationships between the characteristics not readily detectable through human senses or understanding by examining data continuously over periods to detect evidence of trends and potential pending events. Such capability is appropriate in preventing this accident since the cargo paperwork called for maintaining tank temperature below 30°C. The tank reached 50° C three days before the event, 65° C the day before, then rapidly passed 90° C shortly before the explosion. The report identifies a similar spontaneous heating incident in August 2019 affecting a styrene cargo aboard the tanker *Stolt Focus*. In this instance, the crew added seawater to the cargo, then diluted it with benzene and added an inhibitor to bring the situation under control.

4.4.2 *Explosion and Fire*

The Aframax-class crude oil tanker MT *Pablo*, sailing under the flag of Gabon, exploded and caught fire off the coast of southern Malaysia on 1 May 2023 while en route to Singapore with 28 crew members.[34,35] Twenty-five crewmembers were rescued, and three were killed. Built in 1997 and capable of carrying 700,000 barrels of crude oil, the ship was reported to be nearly empty after offloading its cargo in China. Bloomberg reports the

Marshall Islands-registered company that owns the ship does not own any other ships, and no trace of insurance could be found. This highlights the risk of a growing dark fleet of aging vessels transporting sanctioned oil as the result of purchases of hundreds of old tankers by undisclosed buyers.

The cause of the fire is unknown, but since it appears to have been nearly empty, there is a great possibility that vapors from the remains of the oil cargo played a part in this accident. However, the conditions leading up to this accident likely relate directly to the use of old ships subject to the most pervasive types of structural problems of corrosion and fatigue cracking known to be major threats to the structural integrity of aging vessels, especially tanker structures and bulk carriers that operate beyond their design service life.[36] While modern sensor technology may have detected concentrations of explosive vapors before the explosion, their installation and use on an old crude oil tanker is not very probable nor likely to make them safer.

4.4.3 Another Explosion and Fire

An explosion occurred on 1 July 2023 in the tanker *Sagar Nandini-2* while anchored in the river Sugandha near Jhalakathi, Bangladesh.[37] According to the report, the ship was laden with gasoline (petrol) and diesel at the Padma Oil Depot in Jhalakathi town when the engine room exploded, followed by a fierce fire that engulfed the vessel. It was extinguished before it spread beyond the engine room, and no damage occurred to the oil since the chamber containing the oil did not catch fire.

While the exact cause of this accident had not been reported, the spontaneity of the explosion and the fact that the vessel was at anchor suggests several possibilities. These include the ignition of fuel vapors from empty tanks or flammable gas such as methane or hydrogen due to lack of proper ventilation, electrical malfunction leading to sparks or ignition of nearby combustible materials with subsequent fuel or gas ignition, chemical reactions caused by reactive incompatibilities and coming into contact or being improperly stored; and human error or sabotage.

The likely causes of the explosion and ensuing fire initially appear to be unrelated to the ship's function as an oil tanker and might have occurred on any ship. Even so, the explosion suggests that the accumulation of fuel or combustible gas would be capable of being detected using one or more correctly installed and placed smart sensors before igniting a spark. The elements that might cause an explosion were not detected, suggesting a lack of operating sensors for such conditions or human error resulting from ignoring sensor indications and warnings. Sabotage may also be a possibility where human activity sensors and cameras, if installed, could have detected the presence of one or more intruders and recorded their actions or the existence of one or more unauthorized devices within the engine room before the explosion.

4.4.4 Combustion Caused by Welding

In many instances concentrations of leaking gases, liquids and combustible dust may appear at unexpected locations across the ship where fixed monitoring equipment is not installed or has become deactivated through routine maintenance procedures where gases released during loading, unloading, or cleaning may go undetected.

One such case is an accident where the oil tanker *Smooth Sea 22* exploded on 17 January 2023 while docked on the Mae Klong River in the Muang district of Thailand.[38] The explosion killed one welder and left four welders injured. As in the previous example, the cargo may or may not have directly participated in the explosion. However, the crewmembers, temporary workers performing the welding and the ship itself were all placed at risk. In such cases, it may prove extremely valuable to scan areas with portable sensors before work commencement and where workers are present to temporarily install smart sensors capable of detecting explosive dust and vapors that may have accumulated outside of their normal confines.

4.5 Ro-Ro Ships

Ro-Ro carriers are subject to two threats in particular. One threat, Parametric Rolling Movement (PRM), is shared with container ships and is described in a preceding paragraph. The other major threat is fire onboard Ro-Ro carriers that transport vehicles fueled by highly combustible gasoline (petrol) or somewhat less combustible diesel fuel. This issue is highly problematic with these vessels, where the compact loading of cargo described as "a worst-case parking garage" makes it easy for the fire to spread, and firefighting becomes much more difficult.[39] Added to the equation is the diversity and types of cargo, which make it hard to determine what type of fire needs to be fought, and openings on the decks that make detection and localization difficult. Vehicles are stowed 10–15 cm apart during voyages, making access challenging. Typically, fire can spread between vehicles within an alarming five to ten minutes, heightening the importance of identifying the fire's location as early as possible.[40]

The names of some of these car carrier ships are familiar as they have been prominent in the news:

- *Grande Costa d'Avoria* when in July 2023, while completing used vehicle loading operations in Newark, New Jersey, USA, a fire started on deck number ten and burned for more than six days.[41]
- *Höegh Xiamen* burned for eight days in June 2020, started by a fire caused by an electrical fault from an improperly disconnected battery in a used vehicle while in the harbor in Jacksonville, Florida, USA.[42]

- *Grande America* when in March 2019, it burned and sank in the Bay of Biscay off France when a container that was part of the cargo caught fire.[43]

The list of names continues with *Euroferry Olympia* (2022), *Felicity Ace* (2022), *Eurocargo Trieste* (2019), *Grande Europa* (2019), *Golden Ray* (2019: capsized) and several others.

Conventional smoke and heat fire detectors require high threshold levels to trigger a warning and are not sensitive or accurate enough to detect slight thermal changes. Smoke may be rerouted away from smoke detectors by onboard extractor fans, making the detection harder.[42] At present, cargo holds are routinely inspected by crewmembers using predefined walkways between the vehicles to supplement fire detection systems. This is a lengthy and manual process and not always effective, especially when fire risks are not always visible to the human eye. Using infrared cameras to detect heat signatures may help in this endeavor.

One effort currently underway to help resolve these problems is the Swedish Legislative Assessment for Safety Hazards of Fire and Innovations in Ro-Ro Ship Environment (Lash Fire). This initiative is intended to investigate these problems and provide a recognized technical basis for revising international IMO regulations to greatly enhance fire prevention and ensure independent management of fires on Ro-Ro ships in current and future fire safety challenges.[44]

4.5.1 Fires Caused by Lithium-Ion Batteries

A new threat has emerged due to transporting Tesla and many other electric vehicles powered by lithium-ion batteries that are well known to self-combust from thermal runaway and result in extremely difficult-to-extinguish fires. Statistics from the U.S. Bureau of Transportation Statistics and the U.S. National Transportation Safety Board found that hybrid vehicles had the most fires per 100,000 vehicles at 3,474.5; gasoline vehicles were next with 1,529.9 fires per 100,000 vehicles and fully electric vehicles next with 25.1 fires per 100,000 vehicles.[45,46,47] However, electric vehicles can require a tremendous amount of effort from emergency personnel to extinguish the blaze. Lithium-ion batteries also burn hotter and much longer than gasoline and may require tens of thousands of gallons of water to extinguish.

The Safetytech Accelerator challenge was performed in partnership with Wallenius Wilhelmsen, one of the world's largest global vehicle carrier operators.[42] This initiative sought sensor solutions to provide early, real-time detection of significant thermal increases associated with transported vehicles with few or no false positives. The solution must also identify the exact location of the heat change on a ship's deck and accurately communicate the change in real-time through a warning panel. A three-day pilot project provided insight into how a sensor network can be deployed and maintained with reliable results during harsh sea conditions. A Ro-Ro vessel with a typical deck area between 55,000 and 65,000 square meters for cargo would require about 40,000 sensors to provide complete coverage on a ship using a wireless mesh network. Encasing the sensors in robust packaging for the marine environment and fine-tuning their accuracy would cost around 1 USD per node to install, excluding maintenance costs. Costs could be minimized using five-year batteries in the temperature sensor nodes, assuming a typical ship service interval of around five years. Further thermodynamic tests that simulate a vehicle fire more realistically will need to be performed before this technology can be deployed across the fleet.

However, the threat of lithium-ion batteries is not limited to within cargo spaces. The U.S. National Transportation Safety Board (NTSB) in their report determined a November 2022 fire on the oil tanker *S-Trust* was caused by the thermal runaway of a cell within a handheld radio's lithium-ion battery.[48] The event resulted in extensive damage throughout the bridge, including significant smoke and thermal damage, amounting to $3 million in total.

4.5.2 Refrigerator Unit and Other Fire Sources

Before concern regarding lithium-ion batteries became prominent, the European Maritime Safety Agency (EMSA) FIRESAFE final report concluded that fires originating in the cargo spaces (90%) far outweighed ship equipment (5.3%) and other causes (4.7%).[49] The fleet at risk included in the analysis consisted of 490 ships selected based on criteria as agreed by EMSA. SP Technical Research Institute of Sweden AB, Bureau Veritas, and Stena Rederi AB participated in this study.

One of the most significant contributors to fire onboard Ro-Ro ships has been the refrigerator units (33%) that keep containers cool, followed by vehicle drive trains (24%) and truck cabs (17%). Conclusions determined that in most cases, a fire starts as a small smoldering fire and grows very quickly. Also, the development of fires is highly dependent on ventilation and available fuel, and quick response to fire is the most important safety precaution.

4.5.3 Fuel Spill

Transport Malta released its report on an accident onboard the Ro-Ro ship *Eurocargo Trieste* when it experienced a fire in its engine room, most likely caused by fuel spilling accidentally onto a hot surface.[50] Once started, the fire quickly spread due to the presence of combustible material, leaking heavy fuel oil and lubricating oil around the engine and its bilges.

Additional findings cited in the report included:

- The smoke detectors, although operational, failed to alert the crew of the existence of a fire immediately as they probably had been silenced for a short period of time while the vessel was in port.
- Combustible materials in the form of leaked fuel, leaked oil, braided PVC pipes (to direct the leaks), plastic containers to collect drained oils, oil in the bilges and the vicinity contributed to the propagation of the fire.
- It is highly likely that the leak in the CO_2 system compromised its effectiveness.
- Evidence indicated that the doors to the fuel oil modules and separator rooms were open.
- The delay in stopping the port main engine is likely to have contributed to the fire taking hold.

These findings indicate multiple contributory causes, many of which were human in nature such as using plastic containers to collect leaking oil and silencing smoke detectors, thereby exhibiting neglect of proper safety procedures.

Using sensors to enhance safety and situational awareness can only assist seafarers when the sensor indications are observed and not ignored, as in the case of silencing smoke detectors. However, it appears the leak of CO_2 activated the smoke detectors and compromised the CO_2 fire suppression system, making it less effective. This sequence of events was likely confusing, and mistakes may unknowingly have been made. Had appropriate sensors existed within the CO_2 system, this leak may have been detected before the system needed to suppress the fire.

However, one of the most telling of the findings was:

> The decision-making process of the master would have been very complex, involving at least cues (possibly conflicting), technological data, information from fellow crew members, interpretation of that data and a decision to act, either in one way or another.

This finding supports an argument for using automation and smart sensors to aid decision-making.

4.5.4 Improper Stability

The capsize of the car carrier *Golden Ray* in 2019 while departing the port of Brunswick, Georgia, USA, has been attributed to incorrect calculations about the vessel's stability, according to the U.S. National Transportation Safety Board (NTSB).[51] In the report, the chief officer was cited as making an error entering ballast quantities into the stability calculation program.

This led to his incorrect determination of the vessel's stability and resulted in *Golden Ray* having an insufficient righting arm to counteract the forces developed during a turn. The ship caught fire and was ultimately scrapped in place at the accident site. The NTSB also found that two watertight doors had been left open for almost 2 hours before the accident and that no one on the bridge ensured the doors were closed before departure. Nothing appeared amiss until the ship began to heel rapidly to port during a 68-degree turn to starboard.

Although human error is the cause of this accident, there were no backup procedures in place to identify and correct the error. Smart sensors trained in the behavior patterns of the vessel may have detected subtle differences between actual vessel weight, mass, momentum and acceleration forces from nominal before initiating the turn — but maybe not. Indeed, had simple sensors been in place to detect the open doors, the vessel may not have flooded nearly as rapidly.

4.6 Other Ship Types

Many other types of cargos are carried onboard numerous ships of diverse designs. Their sensor requirements to maintain ship stability and detect other threats to the cargo or ship are common with the previously discussed vessel types. Where unique sensor requirements exist due to the nature of the cargo and resulting ship features, solutions are often devised and invented based on specific needs and applications. However, the same basic principles apply to all vessels.

References

1 Zahra Ahmed, Top 20 World's Largest Container Ships in 2023. Marine Insight. 11 April 2023. www.marineinsight.com/know-more/top-10-worlds-largest-container-ships-in-2019/

2 Rishab Joshi. Top 5 Biggest Bulk Carriers in the World. Marine Insight. 6 June 2022. www.marineinsight.com/types-of-ships/biggest-bulk-carriers/

3 TMP Staff. Top 10 Largest Tanker Ships in the World. The Maritime Post. 3 February 2022. https://themaritimepost.com/2022/02/top-10-largest-tanker-ships-in-the-world/

4 Saurabh Sinha. Top 10 Biggest Ro-Ro Ships in the World. 14 March 2022. www.marineinsight.com/types-of-ships/top-10-biggest-roro-ships-in-the-world/

5 Breakbulk's Top Ten. The Maritime Executive. 10 April 2019. www.maritime-executive.com/article/breakbulk-s-top-ten

6 ECIB. ONE Apus Container Loss: Then and Now. Expeditors Cargo Insurance Brokers 28 January 2021. https://insider.ecibglobal.com/blogs/one-apus-container-loss-then-and-now

7 Andrew Gray. Tackling the Scourge of Container Ship Fires. 13 November, 2020. Safety4Sea. https://safety4sea.com/tackling-the-scourge-of-container-ship-fires/
8 Mike Schuler. ZIM Kingston Cargo Fire Stabilized and Ship Held Overnight. 25 October 2021. https://gcaptain.com/zim-kingston-cargo-fire-stabilized-canadian-coast-guard/
9 ABS. Guide for Fire-Fighting Systems for Cargo Areas of Container Carriers. May 2022. American Bureau of Shipping 7.7 FBC Notation (1 October 2019), p. 13.
10 International Convention for Safe Containers (CSC), 1972. International Maritime Organization. Maritime Safety Committee Resolution MSC.355(92). 1 July 2014.
11 George V. Hulme. RFID Helps to Track Cargo Containers. Information Week. 6 April 2005. www.informationweek.com/software/rfid-helps-to-track-cargo-containers
12 Martin LeMonica. IBM and Maresk to Equip Cargo with Sensors. 20 September 2005. CNET.com. www.cnet.com/culture/ibm-maersk-to-equip-ship-cargo-with-sensors/
13 Allianz. Industry's Struggle with Container Ship Fires Continues. August 2018. Safety and Shipping Review. www.agcs.allianz.com/news-and-insights/expert-risk-articles/container-ship-fires.html
14 ING, 2021. Sensor Ships: Why Smart Containers Are the Future of Shipping. Environmental Finance. 1 November 2021. www.environmental-finance.com/content/focus/sustainable-transformation-progress-not-perfection/articles/sensor-ships-why-smart-containers-are-the-future-of-shipping.html
15 Allianz. Industry's Struggle with Container Ship Fires Continues. Allianz Global Corporate & Specialty. August 2018. www.agcs.allianz.com/news-and-insights/expert-risk-articles/container-ship-fires.html
16 Capt. Mark Wood. Parametric Rolling: What It Is and How to Avoid It? Maritime Executive. 28 April 2022. https://themaritimepost.com/2022/04/parametric-rolling-what-it-is-and-how-to-avoid-it/
17 Circular. 1228 – Revised Guidance to the Master for Avoiding Dangerous Situations in Adverse Weather and Sea Conditions. 11 January 2007. Annex, Revised Guidance to the Master for Avoiding Dangerous Situations in Adverse Weather and Sea Conditions. International Maritime Organization (IMO).
18 Karan C. What Is Parametric Rolling in Container Ships? Marine Safety. 6 June 2021. www.marineinsight.com/marine-safety/what-is-parametric-rolling-in-container-ships/
19 Datasheet. Inclinometer Type KPM 100. Kjærulf Pedersen a/s. www.kp-as.com/wp-content/uploads/KPM100-datasheet.pdf
20 Sensor Company Builds System to Warn of Parametric Rolling Danger. The Maritime Executive. 5 September 2022. https://maritime-executive.com/article/sensor-company-builds-system-to-warn-of-parametric-rolling-danger
21 Intercargo. Bulk Carrier Casualty Report, 2019. International Association of Cargo Ship Owners. 1 May 2020. www.intercargo.org/bulk-carrier-casualty-report-2019/

22 Paul Bartlett. Cargo Liquefaction Greatest Cause of Deaths in Dry Bulk Sector. 31 July 2023. www.seatrade-maritime.com/dry-bulk/cargo-liquefaction-greatest-cause-deaths-dry-bulk-sector
23 Susan Gourvenec. The Cargo Ships That 'Liquefy'. BBC. 16 September 2018. www.bbc.com/future/article/20180905-the-cargo-ships-that-liquefy
24 North P&I Club. The Dangers of Cargo Liquefaction in a Nutshell. As described in ShipNerd. 30 May 2022. www.shipnerdnews.com/the-dangers-of-cargo-liquefaction-in-a-nutshell/
25 IMO: Bauxite Liquefaction Sank Bulk Jupiter. Offshore Energy Today. 21 September 2015. www.offshore-energy.biz/imo-bauxite-liquefaction-sank-bulk-jupiter/
26 Gard P&I Club. Cargo Liquefaction. 27 September 2023. www.gard.no/web/topics/article/20651747/cargo-liquefaction
27 Mike Schuler. Gard Warns of Liquefaction Risks with Unlisted Cargoes. gCaptain. 13 July 2023. https://gcaptain.com/gard-warns-of-liquefaction-risks-of-unlisted-cargoes/
28 TMP Staff. Explosion Aboard Chinese Bulker CSSC Cape Town in Gibraltar Bay Leaves Four Injured. The Maritime Post. 22 February 2021. https://themaritimepost.com/2021/02/explosion-aboard-chinese-bulker-cssc-cape-town-in-gibraltar-bay-leaves-four-injured/
29 Clete R. Stephan, P.E. Coal Dust Explosion Hazards. Mine Safety and Health Administration. Pittsburgh, PA, USA. https://ncsp.tamu.edu/reports/MSHA/coaldust.pdf
30 Jiange Chen, Dewen Li, Guoqing Liu, Yanzhu Li, Anran Zhang, Siyuan Lu, and Mi Zhou. Development of a Coal Dust Concentration Sensor Based on the Electrostatic Induction Method. American Chemical Society. ACS Omega 2023, 8, 14, 13059–13067. 28 March 2023. https://doi.org/10.1021/acsomega.3c00319
31 Navy Releases Extensive Bonhomme Richard Fire Report, Major Fires Review. Vice Chief of Naval Operations Public Affairs. 20 October 2021. www.navy.mil/Press-Office/News-Stories/Article/2816283/navy-releases-extensive-bonhomme-richard-fire-report-major-fires-review/
32 USNI. Navy Investigation into USS Bonhomme Richard Fire, Major Fires Review. US Naval Institute. 20 October 2021. https://news.usni.org/2021/10/20/navy-investigation-into-uss-bonhomme-richard-fire-major-fires-review
33 UK Marine Accident Investigation Branch (MAIB) inquiry into the 2019 blast aboard the chemical tanker Stolt Groenland. Report 9/2021. Southampton. Synopsis. July 2021. https://assets.publishing.service.gov.uk/media/60f93e2cd3bf7f044c51590b/2021-09-StoltGroenland-Report.pdf
34 Heather Chen, Irene Nasser and Teele Rebane. Oil Tanker Catches Fire Off Malaysian Coast, Three Crew Missing. CNN. Mon, 1 May 2023. www.cnn.com/2023/05/01/asia/malaysia-coast-oil-tanker-fire-rescue-intl-hnk/index.html
35 Alex Longley and Yongchang Chin. An Oil Tanker Ablaze in the South China Sea Is a Global Problem. Bloomberg. 6 May 2023. www.bloomberg.com/news/articles/2023-05-07/an-oil-tanker-ablaze-in-the-south-china-sea-is-a-global-problem

36 Unyime O. Akpan, T.S. Koko, B. Ayyub and T.E. Dunbar. (2002) Risk assessment of aging ship hull structures in the presence of corrosion and fatigue. Marine Structures, 15(3), 211–231. ISSN 0951-8339. https://doi.org/10.1016/S0951-8339(01)00030-2

37 TBS. Explosion on Oil Tanker in Jhalakathi Leaves Five Burnt, Four Missing. *The Business Standard*. 1 July 2023. www.tbsnews.net/bangladesh/oil-tanker-catches-fire-after-explosion-jhalakathi-5-burnt-4-missing-658470

38 Marine Insight. Oil Tanker Explodes in Thai Waters. Marine Insight News Network. 18 January 2023. www.marineinsight.com/videos/video-oil-tanker-explodes-in-thai-waters-blows-up-workers-leg-500-m-away/

39 Franz Evegren. Fire in Vehicles Onboard Ships. RISE Fire Safe Transport. Research Institutes of Sweden. www.ri.se/sites/default/files/2020-12/FIVE_Fire in vehicles onboard ships_Evegren_WEB_201215.pdf

40 Safetytech. Startup Deploys Wireless Sensors across Ship's Cargo Hold to Predict Fire. Safetytech Accelerator, London, UK. 3 March 2020. https://safetytechaccelerator.org/case-studies/safetytech-startup-deploys-wireless-sensors-across-ships-cargo-hold-to-predict-fire/

41 Mikhail Voytenko. Grimaldi's Ro-Ro Major Fire at Newark, 2 Firefighters Died. FleetMon. 6 July 2023. www.fleetmon.com/maritime-news/2023/42341/grimaldis-ro-ro-major-fire-newark-2-firefighters-d/

42 NTSB. Failure to Properly Disconnect and Secure Vehicle Batteries Led to Fire Aboard Vehicle Carrier Höegh Xiamen. US National Transportation Safety Board. 16 December 2021. www.ntsb.gov/news/press-releases/Pages/mr20211216.aspx

43 FreightWaves Staff. Grimaldi Confirm Grande America Fire Started in Container Cargo. FreightWaves. www.freightwaves.com/news/maritime/grimaldi-confirms-grande-america-fire-started-in-cargo-container-ezkx7-g33nr-gn3cf

44 Legislative Assessment for Safety Hazards of Fire and Innovations in Ro-Ro Ship Environment (Lash Fire). www.lashfire.eu

45 Chris Teague. How Much Should You Worry About EV Fires? Autoweek News. October 2022. www.autoweek.com/news/a38225037/how-much-you-should-worry-about-ev-fires/

46 Traffic Safety Facts Annual Report Tables. National Highway Traffic Safety Administration. U.S. Department of Transportation. https://cdan.dot.gov/tsftables/tsfar.htm#

47 NTSB. Safety Risks to Emergency Responders from Lithium-Ion Battery Fires in Electric Vehicles. U.S. National Transportation Safety Board. www.ntsb.gov/safety/safety-studies/Pages/HWY19SP002.aspx

48 National Transportation Safety Board. Fire aboard Tank Vessel S-Trust, Report MIR-23-23. 25 October 2023. www.ntsb.gov/investigations/AccidentReports/Reports/MIR2323.pdf

49 J. Wikman, F. Evegren, M. Rahm, J. Leroux, A. Bruillard., M. Kjellberg, L. Gustin, and F. Efraimsson. FIRESAFE: Study Investigating Cost Effective Measures for Reducing the Risk from Fires on Ro-Ro Passenger Ships. Lisbon: European Maritime Safety Agency. 2016.

50 Safety Investigation into the Engine-Room Fire on Board the Maltese Registered Ro-Ro Cargo Vessel Eurocargo Trieste. Transport Malta. Marine Safety Investigation Unit. Marine Safety Investigation Report No. 21/2020. Final. www.iims.org.uk/wp-content/uploads/2020/11/Transport-Malta-MV-Eurocargo-Trieste-Safety-Investigation-Report.pdf

51 NTSB. NTSB Determines Inaccurate Stability Calculations Caused Capsizing of Vehicle Carrier Golden Ray. U.S. National Transportation Safety Board. 14 September 2021. www.ntsb.gov/news/press-releases/Pages/NR20210914b.aspx

Chapter 5

Crewmember and Passenger Sensors

Probably one of the most significant areas in which advances in sensor technology have taken place is the result of increasing interest by the shipping industry in the instrumentation of crew and passenger spaces with sensors that not only identify personnel but also track and monitor their movements and well-being while in port and throughout their voyages. This is being accomplished due in part to digitalization in maritime shipping through Port Community Systems (PCSs) and other means to connect different government, public and private organizations for the secure and intelligent exchange of information.

5.1 A Burst of Non-traditional Ship Sensors

The variety and diverse requirements of sensors are as wide-ranging as the passengers and crewmembers that ships carry, the personnel who maintain shipboard systems and others who may or may not have legitimate business onboard. Company-provided crewmember credentials have been supplemented with government-issued identification cards with biometric features allowing access to port and ship infrastructure. Once aboard, additional sensors are used that further verify identity. Likewise, passenger credentialling has changed and now contains vastly greater details to enhance identification, provide access to services and products, and reduce the risk of fraud and unauthorized access to vessels. Long past are the days when guests could be entertained in passenger staterooms before the ship's departure.

Cruise ships integrate fire and smoke detectors into voyager cabins that can now be supplemented with floor, wall and shower sensors to detect occupancy, movement, and even the occurrence of accidents such as slips and falls by the elderly. On some vessels, facial recognition is used to open cabin doors, and wearable devices containing passenger-unique identity information are read by thousands of sensors on the ship and in port to enable hospitality and victualling services and speed the boarding and debarkation processes. These developments are also being adapted to meet the needs of merchant mariners and ship crews.

DOI: 10.1201/9781003377900-5

Data from personnel sensors can be interpreted in combination with ship engineering and navigation sensor data to form a more complete and comprehensive picture to enhance situational awareness and overall vessel and human safety. This chapter describes many of the different types of sensors that can be found on ships and their methods to monitor human activity. Note that some of these methods are complementary in nature and capability, and using more than a few different sensor types on the same ship is highly unlikely.

5.2 Identification Cards

Many countries have developed their own versions of port security card programs of standardized identification systems for workers in sensitive transportation-related industries to help ensure that only authorized individuals with proper credentials can enter restricted areas and carry out their work duties. The sensors and electronic devices that implement these systems and the credentials they examine are identified by different names such as Transportation Worker Identification Card (TWIC), Marine Transportation Security Clearance (MTSC), Maritime Security Identification Card (MSIC), Port Security Identification Card (PSIC) and Maritime Security (MARSEC) Card. However, their purpose remains the same: to enhance security while allowing access to certain restricted areas within ports and harbors, including the vessels within them. The information contained on and encoded in identification (ID) cards generally include the cardholder's full name, their color photograph, date of birth, card expiration date, a unique serial number, card issue date and various security features such as holograms, watermarks and encrypted information containing biometric data including fingerprints and other embedded elements to prevent counterfeiting and ensure authenticity.

Passengers on cruise ships are also provided with identification cards to verify individual personal identities when embarking, disembarking, or reboarding the ship at different ports of call. These cards allow access to the vessel and are used to open their cabin and stateroom doors and to gain access to the many services and facilities on the ship. Specific information contained on and within the ID card vary depending on the cruise line and their systems. In general, it contains the passenger's name and photograph, assigned cabin or stateroom number, cruise ship name or logo and expiration date corresponding to the duration of the voyage. A unique passenger identification number is also assigned, reflecting booking information and onboard accounts to enable purchases such as drinks, dining, entertainment, ship and shore tours, spa services and merchandise. Charges are recorded against the account registered to the card, with the total bill settled by the passenger at the end of the cruise. The card also assists in accounting for passengers during emergency procedures. Large numbers of ID card sensors can exist on individual ships, extending to literally thousands on cruise ships that can track movements and activities throughout the ship on a continuous, round-the-clock basis over spans of days, weeks and even months. Radio Frequency Identification (RFID) technology may also be integrated into ID cards to enable data transfer by passing close to a sensor without requiring contact.

While professional crewmembers, maintenance workers and passengers aboard most vessels must comply with government-mandated identification card requirements, notable exceptions exist on ferries and other relatively short-term transports where such practices are not logistically justified or economically feasible. However, in many cases, the vehicle driver and individual walk-on passengers may be required to show government-issued identification before boarding.

5.3 Closed Circuit Television (CCTV)

Video monitoring has been prominent in previous chapters for surveillance and maintaining security and safety in the engine room, engineering spaces, gangways and other sensitive areas including various entrances, exits, corridors, decks, machinery and storage areas. For navigation, video cameras are used for collision avoidance, aid to navigation (AtoN) detection, visual piloting and maneuvering while docking. They are also used for monitoring cargo spaces, verifying hatch, vent and door closure; and maintaining security for the bridge, the entire ship and its perimeter. It is, therefore, not surprising that video monitoring's greatest role is observing and tracking human activity throughout the ship.

Merchant ships with 20 or so crewmembers are dwarfed compared to the gigantic cruise ships currently being placed into service, capable of carrying over 8,000 crew and passengers. These ships feature 20 or more decks containing many restaurants and lounges, retail stores, child care facilities, swimming pools, rock climbing walls, roller coasters, waterparks, casinos, entertainment forums and other amenities; in addition to machinery spaces, ship's stores, food preparation; waste treatment, disposal and refuse handling; and other functions and services necessary to accommodate the needs of a small city.[1]

Traditional CCTV systems help monitor for suspicious activity such as people loitering, to deter crime and to investigate accidents or disputes by enabling the analysis of the sequence of events, identifying individuals involved and determining the cause of an incident. It can also be used to monitor for safety hazards such as people in unauthorized and high-risk areas like the deck during a storm and people not following proper safety procedures related to the wearing and use of personal protective equipment (PPE) and safety equipment, hard hats, or life jackets.[2,3] CCTV can also be instrumental in demonstrating compliance with and detecting violations of environmental regulations, monitoring lifeboat and fire drills, and for training purposes. However, given the large numbers of people on cruise ships, one of its most important uses is the monitoring of crowded areas like swimming pool decks, dining halls, theaters, or entertainment venues to observe crowd density, detect potential issues before they unfold or help stem their effects, and ensure the smooth flow of passengers, especially during emergencies or evacuation. Many such cameras also have embedded microphones capable of recording human conversations and other sounds in the immediate vicinity and allowing two-way conversations to be carried on through the camera.

Introducing smart video cameras into the ship environment with motion detection and embedded intelligence through machine learning and deep-learning AI provides even more capabilities than traditional video cameras.[4] These capabilities include processing images using facial and object recognition techniques to identify specific people, objects and human activities such as walking, talking and running. Video cameras incorporating thermal imaging capabilities can also distinguish healthy crewmembers and passengers from those exhibiting fevers and possibly suffering from communicable diseases. Video footage may also be stored in the Cloud, allowing company officials and other users to access it from anywhere.

5.4 Facial Recognition

Facial recognition is a method whereby software processes imagery to identify a human face through biometrics and maps facial features from a photograph or video for comparison with a database of known faces to find a match.[5] The primary use on ships involves real-time identity verification by comparing the photograph on the crewmember or passenger identity card with an image retrieved from a live camera image database obtained while boarding or disembarking the vessel. However, many other uses onboard ships and in restricted and secure areas of ports and harbors are possible using this technology.

Facial recognition involves establishing a baseline image using a photograph from which facial features are extracted, including eye position and shape of the nose, mouth, jawline and other facial landmarks converted into a digital mathematical representation.[6] Comparisons are made against a database of face image representations obtained from known individuals. A matching process driven by one or a combination of statistical analysis, machine learning, and deep-learning AI techniques is accomplished to a certain threshold of similarity whereby an identity is considered verified as a match, or not verified. Notification to an operator or computer-based system is then generated for further action in providing access, completing a transaction, or other processes.

Facial recognition is also used to enhance security by constantly monitoring individual people within the ship's spaces to rapidly determine whether they are authorized to be in a particular location or not. Should unauthorized personnel be detected, such as a passenger entering crew-only spaces, an immediate alarm can be raised identifying the intruder and location at which the intrusion took place so that corrective action may be taken. Further use can be made to track movement and activity around a ship to identify locations visited and the times visits occurred, which is especially useful for determining the popularity of cruise ship venues and for tracking an individual's movements while investigating missing person reports on ships.

The accuracy of facial recognition has improved dramatically in recent years. According to data from the U.S. National Institute of Standards and Technology (NIST), the top 150 algorithms are over 99% accurate across black male, white male, black female and white female demographics.[7] For the top 20 algorithms, the accuracy of the highest-performing demographic versus the lowest varies between 99.7% and 99.8%. Unexpectedly, white male is the lowest-performing of the four demographic groups for the top 20 algorithms. For 17 of these algorithms, accuracy for a white female, black male and black female are nearly identical at 99.8%, while they are least accurate for the white male demographic at 99.7%. Detailed statistics for other demographics were not easily available. Facial recognition may be stymied by face masks and plastic surgery. However, this technology appears resilient and can still discern unique patterns sufficient to positively identify people using partial patterns.

5.5 Radio Frequency Identification (RFID)

RFID technology can be incorporated into individual crewmember and passenger ID cards and other wearable devices to supplement printed data as a further precaution to enhance physical and cybersecurity and to help prevent fraud and ID card counterfeiting.[8] This technology operates by storing, and hopefully encrypting, not more than a few thousand bytes of data containing identification numbers, personal information, and access control data and other information within an integrated circuit (IC, or chip) residing in the ID card along with a small antenna that can receive queries and respond by transmitting the data it contains. When an ID card is brought near the RFID reader the reader transmits electromagnetic field emissions that induce a current in a passive antenna providing power to the ID card chip, which then replies by sending the encoded data back to the reader.

RFID technology supports various uses, including granting or denying access to restricted areas, tracking crewmember and passenger movements throughout the ship, and accounting for working hours. In addition, RFID can be used to track the movement of inventory and the location of assets on the ship.

ID cards with embedded RFID technology offer several advantages over traditional identification cards since they are more difficult to counterfeit and easier to use and can be used for a wider variety of applications than traditional ID cards. Disadvantages include their ability to be hacked to retrieve the information they contain, but this vulnerability can be reduced through data encryption. They can also be more expensive than traditional ID cards.

RFID operating frequencies cover three different bands: Low Frequencies (LF), ranging from 30 to 300 kHz; High Frequency (HF), ranging from 3 to 30 MHz, with 13.56 MHz as the dominant frequency; and Ultrahigh Frequency (UHF), ranging from 300 MHz to 3 GHz.[9] The transmission range for passive devices is typically one to five meters or more. Actual ranges will vary depending on the specific tag, reader and environment.

The cost of passive RFID technology continues to decrease, and their utility is increasing with greater memory capacity and two-way read/write capabilities, tending more toward multisurface UHF systems. One study in the proper application of this technology demonstrated the test results of various multisurface UHF systems from different manufacturers for their readability under varying conditions, such as orientation with respect to the reader, distance from the reader and materials used for embedding them.[10] Of ten different products tested, none scored 90% or higher for either readability or precision, so there is ample opportunity to continue enhancing this technology. Antenna orientation and distance to the reader appear to be the greatest contributors to error and inaccuracy. However, some tags seemed to place well on every metric given.

During the period between 2010 and 2020, many enhancements to RFID technology have taken place to improve their performance and accuracy.[11] This includes increased reader sensitivity and read rate, increased transmit power to enhance range, improved processing power with some readers having built-in microprocessors, better antenna designs and additional connectivity options for transferring data.

5.6 Eye Scanning

Optical scanners developed specially to take an image of the eye's retina and the iris can be found on ships to grant or deny access to secure areas. Neither glasses, contact lenses, nor changes in ambient light conditions appear to affect the accuracy or viability of these methods. Scanning of the retina, which is the light-sensitive tissue at the back of the eye, is a biometric technique similar to facial recognition that uses the unique patterns formed by the blood vessels to identify them. The retina and the patterns of its blood vessels are unique to each individual, and such that the patterns they form are often so complex that they are different even between identical twins.[12]

The process used involves a person looking into a scanner that emits a beam of low-intensity infrared light into the user's eye and traces a circular path on the retina. A high-resolution camera records the reflected light to create an image that is then processed to form a digital mathematical model representing a unique identifier for that individual. Comparisons are made against a database of retina models obtained from known individuals using a matching process driven by one or a combination of statistical analysis, machine learning, and deep-learning AI techniques to a certain threshold of similarity whereby an identity is considered verified as a match or not verified.

A study by Bayometric, a global supplier of biometric security systems, found that retinal scans had no false matches while comparing over two million retinal maps. In contrast, fingerprint scanners had more false matches and a lower success rate overall.[13]

Scientists have conducted several studies showing that the human retina can change over time for various reasons, including disease and injury.[14] A common cause of change can be attributed to type 1 and 2 diabetes (diabetic retinopathy) that can damage and burst blood vessels of the retina and thereby invalidate previously acquired retina scan models. Macular degeneration may also cause similar effects in the retina. Researchers have also used human stem cells to grow new retinal cells to replace those destroyed by other medical conditions, which can also change the characteristics of the retina.[15]

Another type of identity verification uses optical scans of the eye's iris, which is the colored part of the eye that contains muscles that control the pupil or the small black opening that lets light into the eye. These same studies have also found that a healthy iris should remain unchanged throughout a person's lifetime. They claim it is impossible to find two completely identical patterns, even in twins. They also note that operations performed on the eyes, removal of cataracts, or implantation of corneal implants do not change the iris's characteristics. In addition, a blind person can also be identified using the iris of their eye. Still, injury from burns, punctures, or strikes with a blunt object and some medical conditions may change the iris.

Iris scanning uses a device very similar in function to retina scanners but targets different parts of the eye. The same basic techniques are used where low-intensity infrared light is aimed into the user's eye and traced across the iris. A high-resolution camera records the reflected light to create an image of the iris that is processed to form a digital mathematical model that represents a unique identifier for that individual.

5.7 Fingerprints

Fingerprints are also used to verify identity because they are unique to each individual, and this method is one of the earliest forms of biometric sensing for identifying people. No two people have the same fingerprints, not even identical twins. This makes fingerprints a very reliable way to identify someone.

Multiple types of sensors may be used for fingerprint identification, including optical, capacitive and ultrasonic scanners. Once scanned, a fingerprint is converted into a digital mathematical model representing a unique identifier for that individual and then stored in a database. Ultrasonic sensors create a 3D fingerprint image that is more secure and accurate than traditional capacitive fingerprint readers. Fingerprint identification uses pattern recognition techniques similar to facial recognition and eye scanning.

The U.S. National Institute of Standards and Technology (NIST) has evaluated several commercially available contactless fingerprint scanning technologies, allowing users to compare their performance to conventional devices that require physical contact between a person's fingers and the scanner.[16,17] The study results show devices requiring physical contact remain superior to contactless technology at matching scanned prints to images in a database. However, when contactless devices scan multiple fingers on a hand, it improves their performance. Contactless devices that scanned multiple fingers also seldom made "false positive" errors that incorrectly matched one person's print with another's record.

Other factors that may make fingerprint matching less reliable include damage to the finger caused by cuts, burns, or other injuries that have occurred since the baseline database model was established, which may reduce or invalidate potential matches. This may also occur should dirt, grease, or other materials obscure the fingerprint; if it is smudged, blurry, or distorted in any way, or if combined with or overlaps fingerprints from other individuals.

5.8 Body Temperature Sensors

Earlier in this chapter, it was mentioned that video cameras incorporating thermal imaging capabilities can distinguish healthy crewmembers and passengers from those exhibiting fevers. Also, present on ships are contactless infrared temperature sensors that, when aimed at a person's forehead, can instantly measure a person's body temperature. These are generally operated by individuals present on a gangway, in a door opening, or in other locations that can restrict personal movement to one individual at a time.

At present, temperature measurements of individuals are accomplished with little or no notice except when it is obvious that a person aims a sensing device straight at someone. Large-scale, real-time measurements of individuals' temperatures for large groups are only possible using wearable temperature sensors such as those within fitness monitoring and medical devices that can communicate using Wi-Fi or Bluetooth wireless connections. Except possibly for ships with few crewmembers, such monitoring is unlikely to become widespread in the foreseeable future due to cost, logistics and privacy issues.

5.9 Space Occupancy Detection

There are many occasions where it is advantageous to determine whether a person or people occupy a room or other space on a ship. This includes the security of physical spaces, detection of motion within secure areas and cargo holds where people are generally not present, or in a passenger ship stateroom to determine opportune times for housekeeping tasks based on a person's habits. Another use is for energy efficiency to turn off room lights, reduce heating temperatures, or increase cooling temperatures when a space is unoccupied.

CCTV is the most accurate sensor for person detection, which was discussed previously in this chapter. However, other sensors are less intrusive and often just as effective while maintaining privacy. One of the most common is passive infrared (PIR) sensors that detect the heat emitted by people and other warm-blooded animals. In rooms with an abundance of glass or metal may be found where PIR sensors may be less effective, an alternative is the use of microwave-emitting sensors that measure reflected signals to detect the presence of people. Still, another alternative sensor uses ultrasound signals to detect the presence of people, especially in areas such as rooms susceptible to dust or smoke.

5.10 Accident (Slip and Fall) Detection

The maritime environment is a hazardous one at best, where slippery surfaces and rolling motions due to changing sea states can cause the most experienced mariners, let alone passengers, to slip and fall that may result in severe injury or even death.[18] Of those 65 and older, who comprise many passengers on ships, more than one in four people fall each year.[19] Although many such occurrences can be prevented by taking proper precautions beforehand, there is no sure prevention method.

The most useful sensor for detecting a slip or fall is a video camera that happens to be aimed at the victim and is being monitored by an operator when an actual fall occurs. However, the likelihood of such combinations of a camera being in the right place and looking at the right time; and a person viewing the camera at the precise time to observe the fall, understand the significance of what they are looking at, and take appropriate action is minimal. Smart cameras can be trained to detect falls and report independent of human activity through proper training using machine learning and deep-learning AI. Still, such applications are relatively new and untested.

Advances in the senior citizen healthcare industry have led to the development of wearable technology that can assist people in alerting others that a fall has occurred by the victim pressing a button and communicating directly to a response center. An alert can also be accomplished through automated alarm generation without any assistance from the victim if incapacitated. Such capabilities can also be afforded to seafarers through wearable devices that monitor for extreme conditions related to velocity, acceleration and pressure outside of nominal values that may be consistent with a fall.

Types of sensors that can be used to detect falls include accelerometers that measure the sudden change in acceleration, gyroscopes that measure the sudden change in angular velocity, and pressure sensors that measure a sudden increase in pressure that occurs when a person falls. Also included are inertial measurement units (IMUs) that use sensor fusion by combining measurements made by accelerometers, gyroscopes and other sensors to determine the overall motion of a moving object with behavioral characteristics that may indicate a slip or fall has occurred.

5.11 Man Overboard Detection

Many devices are available to assist in the detection of a man overboard (MOB) event and recovery of individual crew members or passengers using a transmitter attached to the individual in the water. These maritime survivor locating devices can be used by both crewmembers and passengers and include two basic types: A personal Automatic Identification System (AIS) beacon (PAB) and a personal locator beacon (PLB).[20]

The PAB transmits the wearer's Global Navigation Satellite System (GNSS) position to any vessels in the area up to a 5-mile range via VHF radio Digital Select Calling (DSC) alarm on 161.975 MHz to AIS receivers and AIS-enabled plotters. However, the PLB is a portable device attached to the wearer that sends a distress signal on 406 MHz to the COSPAS-SARSAT satellite network for use by search and rescue (SAR) teams to locate the person in distress. A PLB does not give a localized AIS signal that a vessel could receive.

The best devices include both AIS and PLB capabilities. Both types of devices designed for maritime use can also feature an integrated high-intensity strobe light to assist rescuers in locating the wearer in the water, even in low light conditions.

Cruise ships with many thousands of crewmembers and passengers would find the cost of equipping each with PAB or PLB devices prohibitive, and doing so would likely result in hundreds of false alarms as untrained passengers could inadvertently trigger these devices. However, the U.S. Cruise Vessel Security and Safety Act of 2010 requires passenger vessels operating in the United States to "integrate technology that can be used for capturing images of passengers or detecting passengers who have fallen overboard, to the extent that such technology is available".[21] The International Standards Organization (ISO) 2020 subsequently developed ISO 21195:2020 that specifies technical requirements for systems designed to detect a person who has gone overboard from a ship which specifically excludes MOB detection systems that require the passengers or crew to wear or carry a device such as a PAB or PLB to trigger a MOB event.[22] Several sensor technologies are envisioned to implement ISO 21195:2020 including Radar, infrared, and video imaging, as well as using a combination of technologies to improve the accuracy of MOB detection.[23] Specific technologies for useful MOB detection depend on several factors, including ship size, operating environment and budget. However, all MOB detection systems must meet the requirements of ISO 21195:2020 to be used on ships.

References

1 Molly Bohannon. The 'Icon of the Seas' Will Soon Be the World's Largest Cruise Ship. *Forbes Magazine*. 12 July 2023. www.forbes.com/sites/mollybohannon/2023/07/12/the-icon-of-the-seas-will-soon-be-the-worlds-largest-cruise-ship---heres-how-much-it-costs-to-get-aboard/

2 The use of CCTV Cameras on Ships. Monarch Group. 27 November 2019. www.monarchglobal.net/post/the-use-of-cctv-cameras-on-ships

3 Staff. More Shipowners Use Video Cameras for Security, Safety and Operations. Professional Mariner. 27 May 2016. https://professionalmariner.com/more-shipowners-use-video-cameras-for-security-safety-and-operations/

4 How Do Smart Cameras Work? Technology Org. 18 July 2019. www.technology.org/2019/07/18/how-do-smart-cameras-work/

5 What Is Facial Recognition? How Facial Recognition Works. Internet of Things. Norton. 20 August 2021. https://us.norton.com/blog/iot/how-facial-recognition-software-works

6 What Is Facial Recognition – The 2023 Ultimate Guide for Facial Recognition Technology Guides & Tips. FaceMe. 14 March 2023. www.cyberlink.com/faceme/insights/articles/204/Facial-Recognition-at-the-Edge-The-Ultimate-Guide

7 Patrick Grother, Mei Ngan, Kayee Hanaoka, Joyce C. Yang, and Austin Hom. Ongoing Face Recognition Vendor Test (FRVT). Information Technology Laboratory, National Institute of Science and Technology (NIST), US Department of Commerce. 16 June 2023. www.nist.gov/programs-projects/face-recognition-vendor-test-frvt-ongoing

8 The Use of RFID for Human Identification. Draft Report. U.S. Department of Homeland Security (DHS), Emerging Applications and Technology Subcommittee. Version 1.0. www.dhs.gov/xlibrary/assets/privacy/privacy_advcom_rpt_rfid_draft.pdf

9 ACD. Understanding RFID and RFID Operating Ranges. Advanced Controls and Distribution. 27 March 2017. https://blog.acdist.com/understanding-rfid-and-rfid-operating-ranges

10 Aldo Minardo, Joshua Bolton, Erick Jones, Raghavendra Kumar Punugu, Ankan Addy, and Samuel Okate. Performance and benchmarking of multisurface UHF RFID tags for readability and reliability. Journal of Sensors. Hindawi. 05 September 2017. https://doi.org/10.1155/2017/3467593

11 Suzanne Smiley. RFID Failed You in the Past? It May Have Improved More Than You Think. 19 May 2020. www.atlasrfidstore.com/rfid-insider/why-try-rfid-again/

12 Alexa Saul. Retinal Security Scans: How Accurate Are They? Arizona Retina Project. 31 October 2018. https://azretina.sites.arizona.edu/index.php/node/379

13 Mary Clark. Iris Recognition Scanners vs. Fingerprint Scanners: Compare and Contrast. Bayometric. www.bayometric.com/iris-recognition-scanners-vs-fingerprint-scanners/

14 What Are Iris and Retina Scanners, and How Do They Work? RecFaces. https://recfaces.com/articles/iris-scanner

15 David Turbert. What Parts of the Eye Can Be Transplanted? 13 January 2022. American Association of Ophthalmometry. www.aao.org/eye-health/treatments/transplantation-eye

16 NIST Study Measures Performance Accuracy of Contactless Fingerprinting Tech. 19 May 2020. www.nist.gov/news-events/news/2020/05/nist-study-measures-performance-accuracy-contactless-fingerprinting-tech

17 John Libert, John Grantham, Bruce Bandini, Kenneth Ko, Shahram Orandi, and Craig Watson. Interoperability Assessment 2019: Contactless-to-Contact Fingerprint Capture. NISTIR 8307. National Institute of Standards and Technology. https://doi.org/10.6028/NIST.IR.8307

18 Maritime Slip and Fall/Trip and Fall Injuries. Maritime Injury Guide. www.maritimeinjuryguide.org/maritime-accidents-injuries/maritime-bodily-injuries/slip-fall-trip-fall/

19 Facts About Falls. U.S. Centers for Disease Control. www.cdc.gov/falls/facts.html

20 Fox Morgan. Best Personal Locator Beacons and AIS Units. Yachting World Magazine. 18 July 2023. www.yachtingworld.com/yachts-and-gear/best-personal-locator-beacons-and-ais-units-top-options-for-boating-137237

21 U.S. Cruise Vessel Security and Safety Act of 2010. Public Law 111–207. 27 July 2010, 124 Stat. 2243. 111th Congress. www.congress.gov/111/plaws/publ207/PLAW-111publ207.pdf

22 ISO 21195:2020. Ships and Marine Technology – Systems for the Detection of Persons While Going Overboard from Ships (Man Overboard Detection). www.iso.org/standard/76051.html

23 Elliot Gardner. Setting International Standards for Man Overboard Systems. Ship Technology. 27 May 2018. www.ship-technology.com/analysis/setting-international-standards-man-overboard-systems/

Chapter 6

Artificial Intelligence in Sensor Systems

Throughout the proceeding chapters the use of artificial intelligence (AI) has featured prominently in discussions on smart sensors and sensor data analytics to detect objects, trends and events of human interest with respect to the roles in which seafarers are engaged. This includes the jobs performed by members of the engineering and deck departments in operating and guiding the ship to ensure cargo is loaded, transported and delivered to its destination safely and efficiently; and the hospitality department to safeguard the comfort and well-being of passengers. Critical to this concept is how the various systems and sensors of the ship itself can aid and enhance seafarer situational awareness to improve job and ship safety, performance and effectiveness. AI can provide significant advantages towards achieving this goal, but the technology is not without controversy.

Some argue that AI is the panacea to overcoming human frailties by eventually replacing highly trained and experienced seafarers with automatons that perform their jobs flawlessly, never get bored or need to take breaks or sleep, can work around the clock and do not need to be paid.[1,2,3] Supporting this position are statistics that human error is estimated to be responsible for between 76% and 94% of marine casualties.[4,5] Others think that AI brings the risk of humans losing control to the automaton within the machinery it operates; that external forces can use AI to infiltrate, manipulate, disable and even hijack shipboard systems; and that AI can become superintelligent and capable of malicious behavior to cause harm to those onboard and imperil the ship.[6,7] The benefits of using AI to alert engineering and navigation watchstanders to the development of significant events, conditions, trends and situations can and have been repeatedly demonstrated in enhancing efficient ship routing and fuel savings, collision avoidance, reducing workload, predicting failures and maintenance actions, and detecting objects and people in the water.[8,9] However, the lack of a clear understanding as to exactly how AI systems (especially deep-learning AI) achieve the conclusions they reach, inherent biases in the methods and datasets used to train them, the lack of metrics to verify their development processes and validate their performance,

DOI: 10.1201/9781003377900-6

and the uncanny ability of hackers to penetrate all levels of security precautions lends credence to doomsayer theories.

Many definitions of the term "artificial intelligence" refer to the ability of a machine to simulate, replicate, or even improve upon intelligent behavior exhibited by various forms of life that range from fruit flies to human beings. Since we often fail miserably in discerning how human beings or even insects think and reach conclusions, the best we can do is to compare the results achieved by an AI-based solution against human results for a given, specific and well-defined task. Apparently, AI can be very good at what it does. Spectacular results have been achieved for many years now by AI gaming applications in outwitting human opponents. ChatGPT has successfully passed academic and qualification examinations for a variety of occupations that include medicine, law, accounting and sommelier.[10,11] Internet searches thus far have not yet revealed published records on the success of AI in passing merchant mariner examinations.

Despite the successes of AI cited in the previous paragraph, there is no assurance that such achievements can be translated from an academic environment directly into practical, reliable and widespread use in the real maritime world in much the same manner that a fresh maritime academy graduate is not yet qualified nor sufficiently experienced to stand watch let alone command a ship. However, much research and experimentation have advanced ship autonomy and navigating unmanned voyages along predetermined routes. As applied to ship sensors, there is no well-defined general methodology to ensure AI-based software has been developed correctly or will properly and adequately perform its intended function. Each situation is evaluated on its own merits based on outcomes rather than general and repeatable scientific methods. The question is, is that enough?

This chapter sheds light on how AI may be utilized in sensor-based shipboard systems, what constitutes AI and how these technologies are developed. Benefits of their use and precautions that should be taken when developing, installing and implementing these systems are also discussed. Finally, assertions are made on how these systems may affect seafarers in helping to perform their jobs and enhance safety, how crewmember jobs may change or be eliminated and how seafarer training practices may better adapt to future situations.

6.1 Smart Sensors

A revolution in engineering sensor design has emerged with AI providing greater awareness of and insight into the properties of fluids, flows, and materials; the dynamics and processes used in their handling and changing characteristics at various states and points in time. Individual sensors can monitor various ship components and machinery in real time, learn nominal performance patterns and detect variations and anomalies in the data that are not necessarily perceived nor detected by humans which can help predict equipment failures. A traditional sensor samples the physical environment and then passes the measurement to a system or operator for analysis and action. However, a smart sensor samples the physical environment and, upon detecting specific inputs, uses built-in computer resources to perform predefined functions and then process and analyze the data before passing the data and analytical results to a user or an automated process.[12] The results may alert appropriate crewmembers in sufficient time for them to initiate recommended and appropriate actions, and some can even perform diagnostics and restorative actions on their own. The aggregation of smart sensor data from multiple systems can greatly contribute toward improving overall vessel efficiency and performance. Such capabilities have changed the face of engineering departments such that most engine rooms and many related functions are completely automated. In-person attendance is generally required only on a sporadic basis and when alarms and exceptional situations occur.

New generations of smart sensors can make measurements with greater precision and at higher resolutions. They are fortified with embedded signal processing capabilities that greatly reduce noise and signal loss, with microprocessors that analyze data characteristics at the sensor itself and then communicate their results and enable adjustments to be made to improve the efficiency and performance of the monitored systems. Sensors used in AI-based navigation systems can analyze historical and real-time data that include weather conditions, sea currents and traffic patterns to help optimize ship routes. The fusion of multiple sensor data such as Radar, Sonar, visible light and infrared cameras can help detect, identify and track close-by objects in the ship's vicinity that pose hazards, including other vessels, off-station buoys or floating debris for early detection and identification of potential collisions or obstacles allowing the ship's crew to take appropriate actions. Shipboard sensors and sensors contained within shipping containers can identify sea states that increase their chances of being lost overboard, and sensors contained within bulk cargo can help determine whether cargo liquefaction is likely to occur that may destabilize ships. On the nefarious side, unsolicited and unexpected deliveries of smart watches containing smart sensors to many U.S. sailors and servicemen (and possibly merchant mariners), when turned on, auto-connect to Wi-Fi and cell phones. Without being prompted, access is gained to many user data, cameras, microphones and tracking information while simultaneously inserting malware into these devices.[13]

Benefits of smart sensors include instances where previously required discrete sensor infrastructure components were installed throughout the ship; they are now frequently encapsulated entirely within the sensor itself, resulting in sensor data transmission path distances reduced from meters across the ship to mere millimeters. Shorter signal path distances result in faster data propagation rates, decreased data processing times and reduced chances of injecting unwanted noise and interference into the data, especially when the entire process occurs within the shielded sensor enclosure. Accompanying the sensor are all required signal processing elements combined with computer-based sensor analytics and communications hardware needed to transfer sensor-derived information and analytical results, rather than just the sensor data itself, but also throughout the vessel itself and even worldwide to remote monitoring and control facilities. Further enhancing smart sensor systems are capabilities for self-calibration and integrated diagnostics that lower maintenance costs and improve system reliability by reducing cabling and connector requirements and simplifying remove and repair actions to replacement of individual modules. Additional efficiencies are achieved by taking advantage of distributed computing at the sensor level, enabling rapid adjustments to processes and adapting to changing conditions as they occur. Sensor miniaturization also provides opportunities for including multi-sensor elements within one enclosure, enabling the measurement of several phenomena with one sensor assembly rather than multiple assemblies.

The greatest benefit of smart sensors is probably achieved by taking full advantage of modern microprocessor computing power to examine sensor data in new ways that provide a greater understanding of the phenomenon being measured. This includes the integration of machine learning and deep-learning AI technologies into the sensors themselves that can discern interrelationships between flows, levels, temperatures, and other measured characteristics that are not readily detectable through human senses or understanding. This can be achieved using simultaneous sensor measurements across different sensors, by examining data from one sensor over periods of time to detect evidence of trends and events contained in the data, and with measurements from multiple sensors over time to discern even broader relationships. Having computing power distributed among multiple sensors can effectively achieve at minimal expense what in previous times could only be accomplished using supercomputers at an extraordinary cost. The following paragraphs provide a discussion of what machine learning and deep-learning AI technologies measure and how AI is integrated into sensors and sensor systems.

6.2 Sensor Data Comprising Multiple Perspectives of the Same Subject

The physical properties of things can take many dimensions, each providing a different perspective of behavior or consistency. From one point of view, sensor data may be considered a sample of all relevant characteristics of a measured phenomenon at any single point in time. These characteristics may represent one or more frequency, amplitude, pulse width, resistance, capacitance, or other electrical properties; or a physical property such as level, height, intensity, temperature, and humidity as converted from electrical properties. Commonly measured characteristics of a phenomenon taken as a snapshot at one point in time include multiple signals of interest at different frequencies, much like the different components of a Radar or Sonar signal. Such measurements are considered in the *frequency domain* and represent all relevant signal properties of the phenomenon being measured at that time. When this same measurement (snapshot) is repeated multiple times during a given time period, this sequence is considered in the *time domain*, whereby changes to the signal as they occur over time may be revealed.

Frequency and/or time domain measurements can also be converted into imagery consisting of multiple thousands of picture elements, or pixels, on a display screen relating to the position, area and size of things. This data are represented in the *spatial domain*, which reflects a translation of a spatial quantity into a pixel representation. This can occur by capturing an image of a scene or object directly onto pixels, each containing an impression of the qualities of a small portion of the overall image. The original scene or object is reconstructed by reproducing the pixel impressions onto a display. Such is the case for digital and infrared cameras and other visual sensors. Direct-to-pixel domain imagery is limited based on the size and resolution of the sensor and can be enhanced using optical magnification and greater numbers of smaller pixels, and by using image filtering and software analytics.

Radar, Sonar, and Lidar images are created using an entirely different process involving one or more transducers (antennas) that transmit and/or receive signals. These signals are subsequently converted into different domain representations. An example of Radar is provided in Figure 6.1, where received waveforms are analyzed in the frequency and time domains (b and c) and processed to create a pixel domain representation (a).

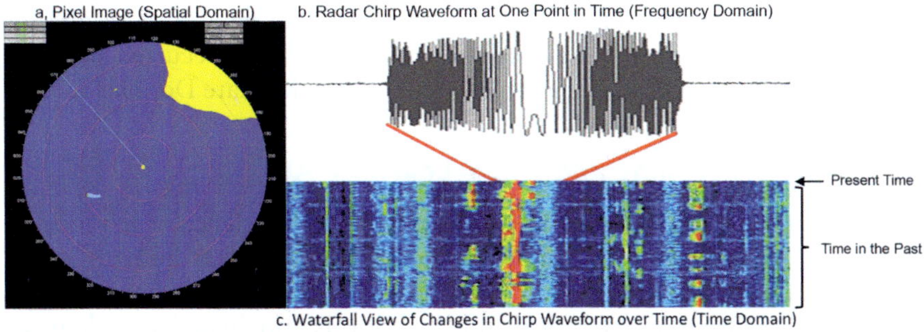

Figure 6.1 Radar Image in Pixels, with Associated Radar Chirp Waveform Variation over Time.

Source: R. Glenn Wright. Intelligent Autonomous Ship Navigation using Multi-Sensor Modalities. 12th International Conference on Marine Navigation and Safety of Sea Transportation. June 2017. Gdynia, Poland.

These electronic signals containing highly complex waveforms across many frequencies are projected through the air or water onto a scene, where the signals are then modified through reflection and absorption based on the physical and electrical characteristics of the environment and the objects present within the scene. A portion of the transmitted signals are reflected to and received by the antenna or transducer, which are then analyzed as a function of changes that occur over time and detected in the frequency spectrum of the signal. Analysis of time and frequency domain signals is performed to acquire the information necessary to subsequently create a pixel domain image for display in a manner unique and customary to Radar, Sonar, and Lidar.

While this indirect approach has proven to be highly accurate and reliable, it can result in great variability in how the targets and scene are displayed to the user based on signal resolution and manufacturer user interface design preferences. For example, a target may be represented as a "blip" on a radar screen, and forward-looking sonar (FLS) can paint a 3D surface model of bottom terrain while Lidar systems can display a highly accurate model of the terrain and quayside environment.

Unlike imaging sensors used in cameras the information contained within the received signal from Radar and similar active sensors in the time and frequency domains creates the resulting pixel domain image based upon the properties of the waveforms being transmitted, the gain and resolution of the transducer elements, the sensitivity of the receiver and the capabilities of the software to analyze the reflected signals. The ability to actively interrogate targets using a wide range of complex waveforms provides great flexibility to analyze their reflected signal properties across all data classes. Dynamic adjustment of waveform signal characteristics in real time based upon target properties and greater capabilities in analyzing time and frequency domain datasets continue to result in the retrieval of much greater information content than has ever been possible. Recent examples include the acquisition of swath bathymetry from FLS and other scientific data from high-resolution side scan sonar imagery.[14,15] Similar advances have also occurred in other maritime applications including improvements in solid-state Doppler radars.

6.3 Machine Learning and Deep-Learning AI

Machine learning is a component of AI and computer science which focuses on the use of data and algorithms to imitate the way that humans learn, gradually improving its accuracy.[16] Traditional algorithms must be written so that their response to every conceivable scenario is anticipated and planned. In contrast, AI is taught or learns on its own the principles involved in a particular application and works out its own responses.[17] The problem is that AI-based systems might respond in unanticipated ways, such as inventing new moves in the game of chess previously unseen by human grandmasters.[18] Much depends on the breadth of the application and the methods and datasets on which the system is trained. It is conceivable that an AI-based system trained to detect and identify navigation buoys could misidentify an oblique view of a small boat painted red with a white number on the side as a navigation buoy.

Various ways to implement machine learning meet the broad definition of allowing computers to learn, including statistical methods, algorithms and neural networks. Statistical methods are based upon mathematical concepts and algorithms in terms of formulas, models, inferencing and other techniques to analyze numerical data and imagery in determining classes, types and variations within and their features and characteristics. Tools to assist in this process include various types of regression techniques, decision trees, and modeling techniques used for predicting and forecasting values, classification, prediction, clustering, and pattern recognition.[19] Depending on the problem's complexity, more sophisticated models and computational architectures may be needed to achieve accurate solutions. This would include artificial neural networks utilizing weighted node structures with multiple visible and hidden layers rather than algorithms. Machine learning techniques do not necessarily require large amounts of training data or long periods of time to produce acceptable results, and the data does not have to be well structured or organized in many cases. Training may be accomplished by mimicking human methods through supervised learning and, in areas where humans are less skilled in performing these tasks, some machine learning architectures can be left to determine trends and outcomes on their own without human direction through unsupervised learning.

Machine learning using statistical and neural network processes can determine similarities between live sensor data and imagery and those corresponding to training data sets to detect and identify trends, events and objects as well as determine sensor performance and possible sensor failure or pending failure. One way to accomplish this is to examine sensor data features and characteristics found outside two and three standard deviations by applying a mask representing qualitative tolerances from nominal values that correspond to nonlinearity, noise, or other aberrations expressed as a percentage of nominal. Other statistical analysis approaches can be followed to reliably and consistently analyze sensor signals considering the ability to detect, classify and identify unique features and characteristics associated with the various data, imagery, waveforms and signal types. Through such means, a wide range of sensors can be supported, providing the greatest applicability across ship and Unmanned Surface Vehicle (USV) remote operations.

The acquisition and analysis of numerical data, imagery, waveforms and other forms of complex sensor signals generally focus on using neural networks designed to meet the specific requirements of each sensor and the objects, events and characteristics of phenomena being sensed. This is especially relevant in cases where complex sensor data exist in the spatial (data that exists in various forms of imagery), frequency (data exist within and across the frequency spectrum) and time (data that change over time) domains defy description in natural language and where numerical and/or algorithmic analysis is burdensome and falls short of comprehensive and complete portrayal of sensor target behavior and characteristics.

Many, even hundreds, of viable neural network architectures may be used for sensor data processing and data analytics, each dedicated to solving specific types of problems and for use in specific applications. One of the more commonly used forms is the Convolutional Neural Network (CNN), capable of detecting numerical data, signal, and imagery features and characteristics needed to identify particular classes, types and variations that may reflect individual objects, trends and events. When using the supervised learning method of training, the relevant features present within the data or imagery are manually extracted and applied in the CNN training process. The advantage of directed learning lies in the reduced time and number of datasets and images needed for training.

Deep-learning AI is also used to train a CNN to detect sensor signal features and characteristics to identify trends, objects and events. In contrast to selecting specific signal and image features and characteristics of interest as is accomplished through machine learning, with deep learning, the raw images are fed directly into a deep neural network where the features are learned automatically. This method generally requires hundreds of thousands to millions of images to achieve the best results, depending on the complexity of the imagery. The advantage of deep learning is in the ability of a CNN to detect hidden relationships present within imagery that may not be readily discernible to someone and prevent their use of the directed learning method. Both methods of neural network training can complement statistical methods for analyzing analog, digital and complex sensor signals and detecting, identifying and categorizing their features, characteristics and attributes.

Another type of deep-learning neural network featured prominently in the news is Generative Adversarial Networks (GAN), where one neural network trained to create imagery (GAN_T) is checked by another trained in the proper boundaries of the imagery (GAN_C). This approach has been used successfully in creating many anatomically correct, "deep fake" human facial images that are highly accurate in their features yet do not represent any actual human beings.[20] Such networks can be used to supplement a variety of actual sensor data for training purposes where the characteristics and constraints of real sensor data are known but difficult to obtain in large volumes.

6.4 Training Dataset Development

Both machine learning and deep-learning AI require large amounts of imagery and numerical data for effective training and solution development. However, these numbers can differ by orders of magnitude, where thousands of datasets are typically used for machine learning. In contrast, hundreds of thousands or more are generally needed for deep-learning AI. Sources for numerical data, imagery and training datasets used for machine learning include the results of computer analysis of sensor signals in the frequency and time domains, combined with spatial domain imagery generated by the sensors and captured by the computer. However, associated risks include amplifying biases against what may be excluded in the training set and favoritism toward characteristics and conditions that are included.

This approach was followed in one project using sensor signal data, waveforms and imagery to generate the sufficiently large datasets needed to assess the potential of deep-learning AI for detecting sensor degradation.[21] The project's original estimate of 100,000+ images rapidly expanded into millions of images and sensor data representations useful for both machine learning and deep learning. Limited datasets were selected, where approximately two-thirds were used for training and one-third for testing, validation, and system evaluation. The training dataset consisted of two parts: a comprehensive dataset and a limited dataset, with the latter being a subset of the former. The comprehensive dataset was shared for supervised learning in the development and refinement of statistical processes and for unsupervised learning in the development of neural network processes. It included complete numerical data providing comprehensive imagery and a description of imagery components that encompass scale, range, units and other factors. The limited dataset was used for directed learning to assist in developing machine learning techniques.

Factors to be considered when developing training datasets for machine learning and deep-learning AI include ensuring the data contains all relevant features and characteristics, and realizing that unsupervised learning will probably require an order of magnitude greater volume than supervised learning. Few, if any, errors should exist in the training dataset if at all avoidable, as this will likely confuse decision-making while in actual use. All features and characteristics should also be normalized to the same scale. Bias in machine learning and deep-learning training datasets where favoring of one class or characteristic of the phenomenon being examined should be minimized or eliminated altogether, when possible. To combat this type of problem, it is necessary to ensure that the training dataset represents real-world conditions in their entire functionality in terms of range, scope and frequency of occurrence.

6.5 Smart Sensor System Verification, Validation and Test

Three aspects of determining sensor system design and operational correctness must be considered throughout the development process, from the initial design through final system acceptance. The first step involves verification that the individual sensors and smart sensor systems fulfill their requirements and have been correctly implemented in accordance with the overall system design, and that interoperability of their individual components when integrated into the entire system is demonstrated. Such a process will identify any missing requirements that should be considered, unnecessary requirements, and that the design process effectively implemented the requirements. This topic becomes difficult with machine learning and deep-learning AI, where unsupervised learning is used when developing the system to define its own requirements as to what is and is not important based on the data presented to it during training. This issue is somewhat less significant when dealing with individual sensors and sensor systems, but becomes vitally important when the concepts of sensor and smart sensor fusion, sensor degradation, and cybersecurity are considered.

Validation of sensor system performance must also be accomplished to determine that the entire system can correctly implement the requirements. The validation process determines whether the requirements contained and presented within training datasets were reasonable in terms of their ability to reflect the actual functions the system must perform, and whether the design was adequate to achieve the desired functions. Problems are likely to arise when smart sensors act upon bias and insufficient scope of data is represented within training datasets to achieve conclusions or solve problems in ways that are correct, or in novel ways that are unanticipated and may or may not be correct.

Testing of sensor system performance must also be performed to cover the breadth and depth of system functionality to determine whether it is sufficiently robust and capable of performing across all required conditions and operational settings. This will include exercising all system functions with a wide variety of test cases to demonstrate its capability to perform its required functions and to ensure that, in the event of failure to perform properly, it can do so in a manner that is not destructive and provides traceability of the cause(s) of failure. The capability of the system to achieve interoperability with other systems must also continue to be demonstrated throughout this process.

A key element to testing smart sensors must consider the possibility of achieving verification and validation of machine learning solutions using statistical processes. However, deep-learning AI solutions are not amenable to this approach due to their inherent lack of visibility into the functioning of their internal processing elements. Merely implementing black box testing techniques based upon system requirements and design is insufficient as decisions are being made in real time, and it may not be possible to effectively reproduce the same outcome by replicating the same exact stimulus recorded by the sensors – a task that is essentially impossible to achieve in an operational environment. This is an area that is still in its infancy, and attempting to achieve a definitive solution to the verification and validation of deep-learning AI is beyond the scope of the present-day state of the art.

Lessons can be learned in the the verification, validation, and testing of deep learning smart sensor applications, by what may be encountered in practice through examining other areas of machine learning and deep-learning AI implementations that have been extensively evaluated and tested on a wide scale by thousands or more users. The problems encountered are indicators of what should be expected to occur using similar technologies.

For example, criticism of the most sophisticated of today's chatbots reveals significant limitations and errors where their results should not be trusted. The OpenAI ChatGPT chatbot developed using deep-learning AI sometimes fails to produce accurate answers, makes up information, uses real organization names to try to legitimize its claims, and even accuses real people of crimes that they did not commit; all the while sounding extremely convincing by producing eloquent prose and citing non-existent sources that seem real even to the people it credits.[22] Other mistakes include getting verifiable facts wrong, detailing impossible processes it cites to create an answer, lying to solve a CAPTCHA test and difficulties in solving simple math problems related to age and distance.[23] Google's Bard chatbot produces errors and false information with high levels of confidence, sometimes offers inaccurate information or omits important details in its answers, and creates results that can inadvertently mirror real-world biases and stereotypes contained in the training datasets.[24]

Based upon these observations and the present state of the technology, AI-augmented sensors and smart sensors are extremely likely to not sense the complete range of objects and conditions for which they are trained, may falsely detect some objects and conditions that do not actually exist, may provide incorrect justification for their findings and just sometimes get things wrong. As such, the use of smart sensors and AI-based decision-making tools should remain relegated to assistant and advisory roles, with trained mariners making all final decisions. Exceptions should be considered in some research and test situations when human safety and lives are not placed in jeopardy, and when significant property damage and environmental risk are not likely to occur.

6.6 Multi-Sensor Data Fusion

The topic of multi-sensor data fusion has prompted much literature over past decades regarding the use of mixed and varied data from multiple sensors to produce a greater awareness and understanding of conditions or events as they occur. Examples can include examining discrete measurements indicating low oil pressure, high oil temperature, low oil flow and fading power to conclude a problem exists where the engine is not receiving a sufficient supply of oil, possibly due to a malfunctioning oil pump. Likewise, a Radar target two miles off the starboard bow at night that cannot be seen visually may be detected by an infrared camera as having the shape of a small boat approaching at fast speed without running lights. What has actually happened here is the fusion of heterogeneous data from different sensors to compare, discriminate and comprehensively analyze the acquired information with memory or experience to obtain a more accurate, complete, and more reliable, consistent interpretation and description of the measured phenomenon than is possible using a single sensor.[25]

Sensor data fusion is a general term that can refer to any method of combining data from multiple sensors. This can include traditional sensors such as cameras, Radar, Sonar, and Lidar; and also smart sensors that can process data locally and make decisions on their own. Much research in this area has been performed using traditional sensors focusing on specific vessel applications related to equipment test, diagnosis and prognosis;[26,27] gaining fuel efficiencies by examining the ship's speed, fuel consumption and weather parameters;[28] and improving object recognition capabilities for autonomous vessel navigation and collision avoidance.[29] The results thus far have been promising and are providing measurable improvements in sensor system designs and gains in efficiency and situational awareness.

However, unlike traditional sensors, smart sensors can communicate with each other and share data, metadata and analytical results in real time, allowing them to collaborate and make better-informed decisions that can aid mariners and directly influence autonomous agents operating unmanned vehicles. Further gains are needed in the design of smart sensors and sensor systems to achieve greater interoperability and for their use to become more commonplace. The fusion of measurement analytical results from multiple smart sensors, not just the measurements themselves, is likely to yield game-changing improvements over traditional sensor fusion through improvements in accuracy and reliability, greater efficiency, enhanced adaptability to challenging situations, increased situational awareness, enhanced decision-making, reduced costs and greater environmental compliance.

6.7 AI and Maritime Training

One bright light for AI use in the maritime industry beyond making sensors better and more useful lies in the realm of training and education.[30] The development of personalized learning methods and tools that can screen for, detect and target areas of knowledge deficiency in a student can measurably improve testing and job performance. The value and realism afforded to training situations and simulations cannot be underestimated when combinations of AI-generated and augmented virtual reality systems can create circumstances where unexpected, real-time curveballs are thrown by an AI-based automaton in adapting training to current situations.

In their recent study on seafarer training and skills needed to support a decarbonized shipping industry, DNV noted a general trend towards a "higher-skilled" seafarer with increased IT, digital, technical and organizational competence being needed in the future and concluded specific training and familiarization onboard are seen as an important part of future training models to ensure seafarers are competent to make use of new technology.[31] Likewise, attracting young people to a career in seafaring is challenging, especially when competing with companies like Microsoft and Google and the working environments they offer.[32] What applies to decarbonization technology and ergonomics in the engineering and deck departments also applies to the understanding and operation of smart sensors onboard ships. Concerted efforts must be made by shipping concerns and training organizations to ensure new generations of smart sensors and sensor systems are understood by crewmembers who may operate and maintain them with the greatest effectiveness.

6.8 Effects of AI on the Maritime Workforce

Introducing AI onboard ships through smart sensors is just one example of how these new technologies are permeating the maritime culture as a whole. Much has been speculated on how AI may affect the jobs and occupations of seafarers, whether automatons will work in harmony with humans to assist them in performing their duties or if seafarers will be replaced. The maritime industry is much like any other in our desire to gain efficiency in operations, and doubtless, the management organizations of maritime firms have noticed many of the impacts AI is having elsewhere. The things AI has demonstrated as doing proficiently include well-structured tasks performed in occupations such as bookkeepers, receptionists, couriers, proofreaders, customer support and technical support.[33] Things AI does not do well include managing people and public relations, entrepreneur skills, human motivation, event planning where many people and contacts are involved, and occupations that require great experience involving many people and complex factors where instincts groomed over many years of experience can be even more valuable than rules and customs in solving problems and making things happen.

The last significant change in the rules for the game of chess occurred in 1849 with the introduction of the Staunton standard to the public, which is still the international standard for tournament play.[34] AI has demonstrated extraordinary abilities to learn, innovate and dominate humans in playing this game over the past several years. Human chess players are still examining and learning new methods and tactics for the opening moves and play introduced by AI.

However, the maritime environment is not static. Indeed, it can change on a minute-to-minute basis by constantly creating new challenges to operate and maintain evolving equipment designs, specifications and performance criteria; and having to cope with new and exciting underway scenarios posed by the effects of weather and influences of the physical environment. Further pressures are exerted on seafarers by having to keep rigid port schedules to ensure timely loading, offloading and replenishment; digesting changes to COLREGS, SOLAS, and other conventions; updates to navigation charts, Notices to Mariners, Coast Pilots and other publications, plus licensing and training requirements for their occupations – and the implications of all these factors for the voyage they are presently conducting and for future voyages. The risks to life, property and the environment are staggering. Despite assertions by many to the contrary, AI cannot begin to cope with so many constantly changing and complex factors. Nor should we expect it to. Unfortunately, regulatory authorities and the automobile and trucking industries have ignored these facts, permitting the introduction of self-driving cars, taxis, trucks and other vehicles onto public highways that have resulted in the slaughter of many innocents and great loss of property.[35,36,37]

Shipowners and maritime regulatory authorities (IMO and nation-states) must not make this same mistake by allowing AI to displace licensed mariners on the bridge and in engineering crews. The maritime workforce is already in decline, and allowing such practices will inevitably accelerate further reduction in properly trained and experienced seafarers. Tracey Mayhew of the Seafarers International Union noted that increased automation will see some jobs become obsolete and some new jobs will be created.[38] The introduction of AI should not exacerbate this problem. The consequences in terms of potential loss of life, environmental damage and high costs incurred through property loss and damage are too high and cannot be justified given the present or near-future states of AI technology maturity. Smart assistants are desirable, but should not be used as an excuse to reduce or replace watch-standers at night or at any other time.[39]

At present, maritime law holds masters responsible for most everything that occurs in or around the ship, including events that are outside their control. However, this policy has its limits. The space shuttle commander was not held liable for the defective voyage planning in 1986 resulting from the near-freezing temperature launch that doomed *Challenger*. It is doubtful that the software and AI industries are prepared to assume liability when the courts determine the automaton "digital navigator" replacing the watch officer makes a mistake that is regarded as negligent navigation which, under admiralty law, does not justify an unseaworthiness claim against the shipowner,[40]

Despite such warnings, there is still a great need for AI in maritime. The primary training center for members of the International Organization of Masters, Mates and Pilots (IOMM&P), the Maritime Institute of Technology and Graduate Studies (MITTAGS), on their website acknowledges the reality that AI is not a serious threat to seafarer jobs:[41]

> As AI increases in popularity on ships and vessels, many crew members and Deckhands are concerned they will lose their jobs as computers can

perform more and more. However, crew members shouldn't worry. While AI and machine learning can streamline and assist with operations, they can only automate so much. Individuals offer unique skills and capabilities that machines cannot replicate, making a human presence on ships invaluable.

This is not to say that seafarers will be unaffected by AI, and this chapter illustrates how AI can help to improve safety by enhancing situational awareness across all seagoing departments. Improvements in seafarer education methods, techniques and facilities to enhance realism and broaden the scope of training can better prepare crew members for the future working with AI to enhance their skills and employment opportunities.

References

1 Eric Niiler. The Robot Ships Are Coming ... Eventually. *Wired Magazine*, Business. 30 October 2020. www.wired.com/story/mayflower-autonomous-ships/
2 Andrew Tunnicliffe. Using AI to Navigate the Tricky Topic of Ship Navigation. *Ship Technology*. 18 April 2019. www.ship-technology.com/features/ship-navigation-system/
3 Annie Brown. Utilizing AI and Big Data to Reduce Costs and Increase Profits in Departments across an Organization. 13 April 2021. *Forbes Magazine*. www.forbes.com/sites/sethmatlins/2023/06/22/the-forbes-worlds-most-influential-cmos-list-2023/
4 Allianz. 2012. Safety and Shipping 1912–2012: From Titanic to Costa Concordia. Allianz Global Corporate & Specialty. 3 March 2012. www/agcs/allianz.com/PDFs/Reports/
5 CBI. 2018. Massive Cargo Ships Are Going Autonomous. *CBInsights*. 28 August 2018. https://app.cbinsights.com/research/autonomous-shipping-trends/
6 Staff. AI: The Worst-Case Scenario. *The Week*. https://theweek.com/artificial-intelligence/1024341/ai-the-worst-case-scenario
7 Matt Egan. Exclusive: 42% of CEOs Say AI Could Destroy Humanity in Five to Ten Years. *CNN Business*. 14 June 2023. www.cnn.com/2023/06/14/business/artificial-intelligence-ceos-warning/index.html
8 Yarden Gross. Why AI Is a First Responder for Overloaded Crews in Shipping's Toughest Era Yet in International Shipping News. 19 August 2021. www.hellenicshippingnews.com/why-ai-is-a-first-responder-for-overloaded-crews-in-shippings-toughest-era-yet/
9 Alexander Love. AI in Shipping: Areas to Watch in 2020. *Ship Technology*. 7 January 2020. www.ship-technology.com/features/ai-in-shipping/
10 Jessica Miley. 11 Times AI Beat Humans at Games, Art, Law and Everything in Between. 12 March 2018. Interesting Engineering, Inc. https://interestingengineering.com/innovation/11-times-ai-beat-humans-at-games-art-law-and-everything-in-between

11 Lakshmi Varanasi. AI Models Like ChatGPT and GPT-4 Are Acing Everything from the Bar Exam to AP Biology. 25 June 2023. www.businessinsider.com/list-here-are-the-exams-chatgpt-has-passed-so-far-2023-1?op=1

12 Brien Posey. What Is a Smart Sensor? *Techtarget.com*. 29 June 2023. www.techtarget.com/iotagenda/definition/smart-sensor

13 NCIS. Free Smartwatches for Sailors Might Be Covert Cyberattack. *Maritime Executive*. 27 June 2023. www.maritime-executive.com/article/ncis-free-smartwatches-for-sailors-might-be-covert-cyberattack

14 FarSounder 2018. FarSounder Joins NOAA as a Trusted Node. FarSounder, Inc. Press Release. 17 October 2018. www/farsounder.com/about/press_releases

15 R. Glenn Wright. Scientific Data Acquisition Using Navigation Sonar. IEEE/MTS Oceans Conference. Anchorage Alaska, USA. September 2017.

16 IBM. What Is Machine Learning? IBM, New York. www.ibm.com/topics/machine-learning

17 Nic Gardner. Current and Future Uses of Artificial Intelligence in the Maritime Industry. Thetius I.Q. 2021. https://thetius.com/current-and-future-uses-of-artificial-intelligence-in-the-maritime-industry/

18 John Nash. Checkmate Humanity: In Four Hours, a Robot Taught Itself Chess, then Beat a Grandmaster with Moves Never Devised in the Game's 1,500-year History. *The Daily Mail*. 21 December 2017. www.dailymail.co.uk/sciencetech/article-5204513/Robot-taught-never-seen-chess-moves-hours.html

19 Coursera. 10 Machine Learning Algorithms to Know in 2023. Coursera, Inc. Updated 16 June 2023. www.coursera.org/articles/machine-learning-algorithms

20 Tero Karras, Samuli Laine, and Timo Aila. *A Style-Based Generator Architecture for Generative Adversarial Networks*. NVIDIA. arXiv:1812.04948v3 [cs.NE], 29 March 2019.

21 R. Glenn Wright. In-Stride Detection of Sensor Degradation. GMATEK, Inc. Final Report. Revision: A, 28 April 2022. Contract: N6833520G2005. Naval Sea System Command, Washington, DC.

22 Emily A. Vogels. A Majority of Americans Have Heard of ChatGPT, but Few Have Tried It Themselves. Pew Research Center. 24 May 2023. www.pewresearch.org/short-reads/2023/05/24/a-majority-of-americans-have-heard-of-chatgpt-but-few-have-tried-it-themselves/

23 ChatGPT Fails: 13 Common Errors and Mistakes You Need to Know. Search Engine Land. Accessed 5 July 2023. https://searchengineland.com/chatgpt-fails-errors-mistakes-400153

24 Google Bard AI Chatbot Advantages, Disadvantages, Review, Features & Bard vs ChatGPT. Online Sciences. Accessed 5 July 2023. www.online-sciences.com/technology/google-bard-ai-chatbot-advantages-disadvantages-review-features-bard-vs-chatgpt/

25 Guo Chen, Zhigui Liu, Guang Yu, and Jianhong Liang. (2021) "A new view of multisensor data fusion: research on generalized fusion," Mathematical Problems in Engineering, 2021, Article ID 5471242, p. 21. https://doi.org/10.1155/2021/5471242

26 Wen Jiang, Weiwei Hu, and Chunhe Xie. (2017). A new engine fault diagnosis method based on multi-sensor data fusion. Applied Sciences, 7(3), 280. https://doi.org/10.3390/app7030280

27 Youyu Zeng and Qiang Xie. Real-Time Ship Fault Diagnosis Algorithm based on the fusion of CNN and RNN. Proc. 2020 the 10th International Workshop on Computer Science and Engineering (WCSE 2020). Shanghai 19–21 June 2020, pp. 70–75. doi: 10.18178/wcse.2020.06.012. ISBN 978-981-14-4787-7

28 Yuquan Du, Yanyu Chen, Xiaohe Li, Alessandro Schönborn, and Zhuo Sun. (2022) Data Fusion and Machine Learning for Ship Fuel Efficiency Modeling: Part III – Sensor Data and Meteorological Data. Communications in Transportation Research, 2, p. 100072, ISSN 2772-4247, https://doi.org/10.1016/j.commtr.2022.100072

29 DNV-Maritime. New Research Project to Investigate Sensor Fusion and Collision Avoidance for Advanced Ships. DNV. Accessed 29 June 2023. www.dnv.com/news/new-research-project-to-investigate-sensor-fusion-and-collision-avoidance-for-advanced-ships-26127

30 P. Roberts. Effects of Artificial Intelligence on Maritime Training. 31 March 2023. *Shoreside Maritime Training Magazine.* https://shoresidetraining.co.uk/news/the-effects-of-artificial-intelligence-on-maritime-training/

31 Raymond Antoni Kaspersen (PM), et al. Insights into Seafarer Training and Skills Needed to Support a Decarbonized Shipping Industry. DNV. Report no. 2022-0814, rev. 0. 2022-11-04. General Conclusions 3 and 4, p. 8.

32 Marcus Hand. Attracting Future Seafarers and Engine Room Ergonomics. *Seatrade Maritime News.* 29 June 2023. www.seatrade-maritime.com/crewing/attracting-future-seafarers-and-engine-room-ergonomics

33 Sophia Bernazzani, 10 Jobs Artificial Intelligence Will Replace (and 10 That Are Safe). Blog Hotspot. Updated: 21 February 2020. https://blog.hubspot.com/marketing/jobs-artificial-intelligence-will-replace

34 The Staunton Standard: Evolution of the Modern Chess Set. World Chess Hall of Fame. St. Louis, MO, USA. https://worldchesshof.org/exhibit/staunton-standard-evolution-modern-chess-set

35 Ariana Bindman. Incompetent Driverless Cars Are Wreaking Havoc on San Francisco. SFGATE, 6 June 2023. www.sfgate.com/tech/article/cruise-waymo-driverless-cars-san-francisco-18132953.php

36 Nearly 400 Car Crashes in 11 Months Involved Automated Tech, Companies Tell Regulators. The Associated Press. 15 June 2022. www.npr.org/2022/06/15/1105252793/nearly-400-car-crashes-in-11-months-involved-automated-tech-companies-tell-regul

37 Ed Leefeldt and Amy Danise. Injured in a Car Accident? AI Is Your Legal Adversary. *Forbes Magazine.* 9 June 2023. www.forbes.com/advisor/car-insurance/car-accident-injury-artificial-intelligence/

38 Tracey Mayhew. Uncertainty for Maritime Workforce of the Future. MARAD Symposium, Achieving Critical MASS. 23 July 2019. www.maritime.dot.gov/sites/marad.dot.gov/files/docs/about-us/foia/11696/seafarers-international-union-uncertainty-maritime-workforce-future.pdf

39 Michael Grey. The Case for Fewer Eyes on the Bridge at Night. *Seatrade Maritime News*. 20 June 2023. www.seatrade-maritime.com/crewing/case-fewer-eyes-bridge-night

40 The Maritime Executive. Will Owners of Autonomous Ships Be Liable for Autonomous Mistakes? 19 May 2023. https://maritime-executive.com/article/will-owners-of-autonomous-ships-be-liable-for-autonomous-mistakes

41 MITAGS. How AI Is Changing the Maritime Industry. Maritime Institute of Technology and Graduate Studies. Linthicum, MD, USA. Blog, News. 10 April 2023. www.mitags.org/ai-impact-maritime-industry/

Chapter 7

Shore and Intership Communications

Electronic and optical technologies provide a means to extend the human senses of vision and hearing and to project our ability to speak from the vessels on which we sail to span the local vicinity and out to the entire world in real time. It also enables our capacity to communicate data and information generated by the electronic sensors installed on our vessels at sea to other vessels, corporate offices, Remote Operating Centers (ROCs), national authorities and other locations where such a need exists. This includes access to telecommunications and broadband infrastructure for crews to contact families and people ashore; and to fulfill many maritime-specific needs to communicate with other ships of any size or nationality related to urgent maritime safety information, distress alerts in an emergency and contact rescue coordination centers anywhere in the world as well as nearby ships.

This chapter describes means that exist to facilitate one-way and two-way communication using modern technological tools that include wireless and physical connections. Vessels must be appropriately equipped, and seafarers must have a working knowledge of many modern tools under the SOLAS Convention. Traditional communication methods involving sight and sound also remain vital for all operations in the immediate vicinity of the vessel through the use of lights, dayshapes, whistles, horns, bells, smoke, fire, and gunfire.

While traditionally staffed and real-time, remotely operated vessels have crewmembers knowledgeable and experienced in the art of communication, unmanned vehicles must be taught not only these methods and techniques but also the meaning and significance of the content of these communications.

7.1 Communications Infrastructure

The single most significant advance in communications to and from ships at sea in the twenty-first century has been the advent of Wide Area Networks (WAN) capable of spanning vast distances and providing many circuit paths

DOI: 10.1201/9781003377900-7

to achieve global connectivity. Shipboard communications equipment utilizing satellite and technologies make it possible to access the Internet using public and dedicated private networks of various and diverse configurations of fiber optic and copper cables, microwave links, ground stations and terrestrial networks. These same infrastructure form the backbone of VHF and HF GMDSS messaging and other national emergency services worldwide. Public Metropolitan Area Networks (MAN) provided by regional governments, local cities and municipalities that serve limited geographic areas such as ports and harbors achieve much the same connectivity using Wi-Fi connections, as do private services associated with ports and industries within the ports limited to ships and other entities by their need to do business.[1]

Detailed discussion on the topic of worldwide communications infrastructure is deferred to others who consider the broader aspects of information technology (IT) as it applies to individuals, businesses and governments in general, where communications by ships comprise only a tiny part of their scope. However, it is mentioned here to provide a background that supports the various means of communication by which ships interact with other ships and shore entities. The following paragraphs discuss these various means of communication by ships.

7.2 Satellite Communication

Satellites are commonly used for a wide variety of navigation-related maritime voice and data communications that include routine status updates between ships and their home offices, tracking ship movements, and increasingly, command and control of conventionally staffed vessels and unmanned and autonomous vehicles. Also included are communications related to the interruption of normal navigation by emergencies with regard to distress signals and critical messages related to accidents, loss of propulsion, medical emergencies, piracy and significant matters requiring immediate attention. This mode of communication is greatly relied upon while at sea for voice and data communications due to its reliability and large bandwidth capacity.

Satellites in Low Earth Orbit (LEO) are located between 100 and 1,200 miles above the Earth's surface, with an orbital period of between 90 and 130 minutes. The operator with the most LEO satellites is SpaceX which, as of 2023, has approval to launch 7,500 of its planned 29,988-spacecraft constellation.[2] Other LEO satellite operators include Starlink, OneWeb, Iridium, Inmarsat/Viasat[3], SateliotX, Orbcomm, and Saab. One notable Medium Earth Orbit (MEO) satellite provider for maritime is SES with their O3b mPower broadband services. Located in Geosynchronous Earth Orbit (GEO) is Inmarsat/Viasat, which provides its Orchestra, Elera, and Global Xpress networks.

A wide range of different bands of frequencies are used for satellite communications depending upon service providers, the types of data transferred and bandwidth requirements. These include the L-band (1–2 GHz), S-band (2–4 GHz), C-band (4-8 GHz), X-band (8–12 GHz), Ku-band (12–18 GHz) and the Ka-band (26.5–40 GHz) frequencies.

7.2.1 *Very-Small-Aperture Terminal (VSAT) Satellite Systems*

Very-Small-Aperture Terminal (VSAT) has gained wide acceptance for maritime communications due to its small antenna footprint that makes installation possible on many smaller vessels at relatively lower cost when compared to more traditional, high-bandwidth satellite systems. These systems provide Internet access as well as voice, video and data communications, fleet management and other applications using satellites positioned in GEO that operate in the C, Ku and Ka frequency bands. To gain additional bandwidth, increase reliability, reduce latency and overcome the limitations associated with any one satellite services technology or provider, many users are combining VSAT with LEO systems such as Starlink and 5G services as a backup to L-band services for guaranteed worldwide connectivity.[4]

7.2.2 *Satellite-Based Automatic Identification System (S-AIS)*

An early version of S-AIS commenced in 2008 with the launch of an Orbcomm concept demonstration satellite equipped with AIS capability under the U.S. Coast Guard Nationwide Automatic Identification System (NAIS) project to test the feasibility and effectiveness of Automatic Identification System (AIS) message reception and reporting from space for ship tracking and other navigational activities.[5] Data collected by the NAIS would support maritime interests by promoting vessel and port safety through collision avoidance and through detection, traffic identification and classification of vessels out to 2,000 nautical miles from shore. Subsequently, S-AIS services have expanded worldwide through other satellite systems and many commercial services to support vessel management to ensure safer ship routing, enable Vessel Traffic Monitoring and Information Systems (VTMIS), promote ship reporting, validate ship declarations, offer efficient Vessel Traffic Services (VTS), manage arrivals and perform effective vessel traffic analysis.[6] S-AIS is also a valuable tool to track and assist vessels in distress, monitor piracy and prevent smuggling, and aid in environmental preservation and fishery protection.

7.2.3 Remote Vessel Control via Satellite

Communication satellites also provide the primary means of remote control navigation of vessels. First demonstrated in 2017, the 28 m tug *Svitzer Hermod* safely conducted a number of maneuvers in Copenhagen harbor while being controlled by an ROC.[7] Wärtsilä also remotely controlled their *Highland Chieftain* platform supply vessel in the North Sea using a standard satellite link over 5,000 miles from its office in San Diego, California.[8] In 2020, the fully autonomous *Mayflower* equipped with 6 smart cameras, 30 onboard sensors and 15 intelligent network edge devices capable of analyzing and using data in real time sent data and metadata from the voyage back to the onshore team via Iridium Certus weather-resilient broadband satellite connections.[9,10] Such technology is steadily working its way into all aspects of commercial shipping as a means for companies to gain further insight into operational parameters while maintaining steady contact with their vessels at sea and to assist in emergency situations should remote control of the vessel be needed.

7.2.4 Limitations of Satellite Communications

Providing a broad expansion of services to vessels at sea comes with limitations associated with satellite communications that should be noted. These can include frequency congestion, interference from weather and sunspots, and propagation delay for GEO satellites. Coverage of specific geographic areas may vary, especially with LEO and MEO satellites that are constantly changing their relative position in the sky with respect to the ground and when another satellite is not available to take the place of a satellite that has gone out of view. Satellite communications are also more vulnerable to security risks than terrestrial communications, as their data can be intercepted by third parties.

7.3 Very High Frequency (VHF) Radio

Maritime VHF-FM radio provides two-way voice and data communications between vessels and shore stations. The useful distance of a maritime VHF radio is generally within line of sight depending upon the height of the antenna, weather conditions, interference from other users and natural events. VHF radios with 25-watt output power can have a range of 20–60 nautical miles between antennas mounted on tall ships, towers and hills; and 5 nautical miles between antennas mounted on small boats at sea level. Small handheld transceivers with 5-watt radiated output power cover considerably less distances. Table 7.1 illustrates a list of VHF international transmitting frequencies.[11]

Useful for distress and safety communications, VHF radio is one of the primary means of communication in the event of a distress or safety situation at sea. Channel 16 is designated as the international distress, safety and calling frequency that vessels are required to monitor at all times while underway. Channel 13 is dedicated for bridge-to-bridge communications in the local vicinity to assist in navigation by coordinating activities such as passing or docking or sharing information about weather conditions or hazards.

Broadcasts containing weather forecasts and other marine information can also be received on this channel. Another use is to communicate with shore stations such as marinas, harbormasters and the Coast Guard; and for requesting assistance, reporting marine incidents, or making arrangements for fuel, food, or other supplies.

There are many restrictions and operationally relevant conditions that are specific and subject to national and regional requirements and regulations not illustrated in this table. Examples include frequency use by land and aircraft stations for search and rescue coordination, pollution and other safety-related operations; radiated power restrictions; digital selective calling for distress, safety and calling; AIS and mobile-satellite service for the reception of AIS transmissions; simplex and duplex frequency transmissions; restrictions from harmful interference; and frequency use for digital systems and new technologies.

7.3.1 Digital Select Calling (DSC)

Initiated as part of the Global Maritime Distress and Safety System (GMDSS), a new generation of marine VHF (and HF) radios feature a Digital Selective Calling (DSC) capability that enables mariners to send or receive distress, urgency, safety and routine radiotelephone calls to or from any similarly equipped vessel or shore station including automatic digitally formatted distress alerts to coast guards or other rescue authorities anywhere in the world.[12] These communications may be encrypted for security purposes. DSC is a mandatory requirement under the SOLAS Convention for all ships that are required to carry a VHF radio, and is recommended for all other ships.

DSC utilizes the Maritime Mobile Service Identity (MMSI) and call sign programed into the VHF radio upon initial installation of the vessel, and position information acquired using a built-in GNSS capability present within the radio itself or obtained from an onboard GNSS received connected to the radio. The MMSI of the recipient vessel or shore station is entered, and the message is initiated by pressing the "send" button on the radio. DSC is an essential GMDSS component to improve safety at sea.

Table 7.1 VHF International Transmitting Frequencies

Channel Designator	*Notes*	*Transmitting Frequencies (MHz)*		*Intership*	*Port Operations and Ship Movement*	
		Ship Stations	*Coast Stations*		*Single Frequency*	*Two Frequency*
60	*m*	156.025	160.625		x	x
1	*m*	156.05	160.65		x	x
1001		156.05	156.05		x	
61	*m*	156.075	160.675		x	x
2	*m*	156.1	160.700		x	x
62	*m*	156.125	160.725		x	x
3	*m*	156.15	160.750		x	x
63	*m*	156.175	160.775		x	x
1063		156.175	156.175		x	
4	*m*	156.2	160.800		x	x
64	*m*	156.225	160.825		x	x
5	*m*	156.25	160.850		x	x
1005		156.25	156.25		x	
65	*m*	156.275	160.875		x	x
1065		156.275	156.275		x	
6	*f*	156.3		x		
2006	*r*	160.900	160.900			
66	*m*	156.325	160.925		x	x
1066		156.325	156.325		x	
7	*m*	156.35	160.950		x	x
1007		156.35	156.35		x	
67	*h*	156.375	156.375	x	x	
8		156.4		x		
68		156.425	156.425		x	
9	*i*	156.45	156.45	x	x	
69		156.475	156.475	x	x	
10	*h,q*	156.5	156.5	x	x	
70	*f,j*	156.525	156.525	Digital selective calling for distress		
11	*q*	156.55	156.55		X	
71		156.575	156.575		X	
12		156.6	156.6		X	
72	*i*	156.625		x		
13	*k*	156.65	156.65	x	x	
73	*h,i*	156.675	156.675	x	x	
14		156.7	156.7		x	
74		156.725	156.725		x	
15	*g*	156.75	156.75	x	x	
75	*n,s*	156.775	156.775		x	
16	*f*	156.8	156.8	Distress, safety and calling		
76	*n,s*	156.825	156.825		x	
17	*g*	156.85	156.85	x	x	
77		156.875		x		
18	*m*	156.9	161.5		x	x
1018		156.9	156.9		x	
78	*m*	156.925	161.525		x	x
1078		156.925	156.925		x	
2078	*mm*	161.525	161.525		x	
19	*m*	156.95	161.55		x	x
1019		156.95	156.95		x	
2019	*mm*	161.55	161.55		x	

Source: US Coast Guard, 2023. VHF International Transmitting Frequencies. US Coast Guard Navigation Center. Department of Homeland Security. www.navcen.uscg.gov/international-vhf-marine-radio-channels-freq

Public Correspondence	*Channel Designator*	*Notes*	*Transmitting Frequencies (MHz)*		*Intership*	*Port Operations and Ship Movement*	
			Ship Stations	*Coast Stations*		*Single Frequency*	*Two Frequency*
x	79	*m*	156.975	161.575		X	x
x	1079		156.975	156.975		X	
	2079	*mm*	161.575	161.575		X	
x	20	*m*	157	161.6		X	x
x	1020		157	157		x	
x	2020	*mm*	161.6	161.6		x	
x	80		157.025	161.625	x	x	
	1080		157.025	157.025		x	
x	21	*y,wa*	157.05	161.65		x	x
	1021		157.05	157.05			x
x	81	*y,wa*	157.075	161.675		x	x
x	1081		157.075	157.075		x	
x	22	*y,wa*	157.1	161.7		x	x
	1022		157.1	157.1		x	
x	82	*x,y,wa*	157.125	161.725		x	x
	1082		157.125	157.125		x	
	23	*x,y,wa*	157.15	161.75		x	x
	1023		157.15	157.15		x	
x	83	*x,y,wa*	157.175	161.775		x	x
	1083		157.175	157.175		x	
x	24	*w,wx,x,*	157.2	161.8		x	x
	1024	*w,wx,x,*	157.2				
	2024	*w,wx,x, xx*	161.8	161.8	x (digi		
	84	*w,wx,x,*	157.225	161.825		x	x
	1084	*w,wx,x, xx*	157.225		x (digi		
	2084	*w,wx,x,*	161.825	161.825			
	25	*w,wx,x,*	157.25	161.85		x	x
	1025	*w,wx,x, xx*	157.25		x (digi		
	2025	*w,wx,x,*	161.85	161.85			
	85	*w,wx,x,*	157.275	161.875		x	x
	1085	*w,wx,x,*	157.275				
	2085	*w,wx,x,*	161.875	161.875			
	26	*w,ww,*	157.3	161.9		x	x
	1026	*w,ww,*	157.3				
	2026	*w,ww,*		161.9			
	86	*w,ww,*	157.325	161.925		x	x
	1086	*w,ww,*	157.325				
	2086	*w,ww,*		161.925			
	27	*z,zx*	157.35	161.95			x
	1027	*z,zz*	157.35	157.35			
x	ASM 2	*z*	161.95	161.95			
	87	*z,zz*	157.375	157.375		x	
x	28	*z,zx*	157.4	162			x
	1028	*z,zz*	157.35	157.35		x	
	ASM 2	*z*	162	162			
x	88	*z,zz*	157.425	157.425		x	
	AIS 1	*f, l, p*	161.975	161.975			
	AIS 2	*f, l, p*	162.025	162.025			

7.3.2 Weather Broadcasts, Notices to Mariners

Many countries broadcast maritime weather information on maritime VHF radio. Some of the most common channels used for this purpose include:

- Channel 16 (156.8 MHz): The international distress, safety and calling channel, it can also be used for safety-related weather information broadcasts.
- Channel 6 (156.3 MHz): Weather broadcasts in the United States, Canada and some other countries.
- Channel 22A (157.1 MHz): Weather broadcasts in the United Kingdom and Ireland.

Broadcast Notice to Mariners (BNMs) provide mariners with information about potential hazards such as dangerous weather, shoals, wrecks and other vessels may also be transmitted on VHF channels 16 and 22A. Different VHF frequencies over which national weather offices may send local and nearshore coastal marine forecasts and weather observations and forecasts include 162.400 MHz, 162.500 MHz, 162.425 MHz, 162.525 MHz, 162.450 MHz, 162.550 MHz and 162.475 MHz. Coastal stations with ranges between 20 and 25 miles offshore can also broadcast predicted tides and real-time observations from buoys and coastal meteorological stations. Specific channels and frequencies used for maritime weather broadcasts can vary from country to country.

7.3.3 Emergency Services

Emergency services are available through maritime VHF radio under programs supported by the United States Coast Guard, the European Maritime Safety Agency (EMSA) and other national entities. The Rescue 21 program is the U.S. Coast Guard's advanced command, control and direction-finding communications system created to better locate mariners in distress and save lives and property at sea and on navigable rivers.[13] Rescue 21 helps identify the location of mariners in distress using VHF radio via towers that generate lines of bearing to the source of radio transmissions, significantly reducing search time, meeting both the safety requirements needed for the expanding U.S. Marine Transportation System and for International Convention for the Safety of Life at Sea (SOLAS) standards. It extends coverage to a minimum of 20 nautical miles from the coastline; improves information sharing and coordination with the U.S. Department of Homeland Security and other federal, state and local first responders; and can also help watchstanders recognize potential hoax calls by identifying discrepancies between a caller reported and actual locations, thus conserving and reducing risk to valuable response resources. Other features of the Rescue 21 system include improved clarity, recording and playback capabilities for all communications, simultaneous monitoring and relaying of information over multiple radio frequencies, portable tower communications equipment for use during emergencies and natural disasters and automated transmission of urgent marine information broadcasts.

The EMSA is an agency of the European Union that is responsible for maritime safety and security. It operates several programs that provide features similar to the U.S. Rescue 21 program, including the European Coastal Monitoring System (ECMS) and the European Maritime Safety Information System (EMSIS). EMSA programs acquire a comprehensive overview of activities at sea, which is a challenge for most authorities working in the maritime domain through their Integrated Maritime Services.[14] They aim to provide coast guards and border guards, customs authorities, naval forces, pollution response centers, rescue coordination centers and vessel traffic management services all need detailed, reliable information about occurrences at sea in real time to perform their duties effectively.

7.3.4 VHF Signal Propagation Effects

A significant exception to the stable and dependable characteristics of VHF radio can be attributed to the phenomena of "atmospheric ducting", where occasionally the transmission of VHF electromagnetic waves through a region near the tropospheric layer of the atmosphere can literally extend hundreds of miles or more beyond the line of sight with little signal attenuation.

Despite normal signal absorption from being reflected from the ionosphere or passing over the surface of the earth, ducting can be attributed to radio waves propagating from one end to the other end of the duct by undergoing successive refraction from the troposphere.[15] This duct is a region or layer present in the lower atmosphere occurring from a temperature inversion that behaves as a waveguide propagating radio waves along the curvature of the earth and allows the spreading of wavefronts in a horizontal manner.

7.4 Medium Frequency (MF) and High Frequency (HF) Radio

Maritime MF and HF radio also provide two-way voice and data communications between ships, boats and shore stations in the same manner as VHF radio, but at much lower frequencies with drastically different radio wave propagation characteristics. The useful distance of maritime MF and HF radios can generally range from less than one hundred miles to thousands of miles depending upon the frequencies involved and several other factors. Tables 7.2 and 7.3 list the MF and HF Single Sideband Maritime Radiotelephone Channels.

Table 7.2 MF Single Sideband Maritime Radiotelephone Channels

2 MHz Simplex Channels	2.003 kHz	2.142 kHz	2.214 kHz	2.738 kHz
	2.0825 kHz	2.182 kHz	2.635 kHz	2.782 kHz
	2.086 kHz	2.203 kHz	2.638 kHz	2.830 kHz
	2.093 kHz			

MF radio waves in the 2 MHz band are characterized by a ground wave that can propagate over the horizon, and skywave propagation that can be reflected by the ionosphere. The ground wave is the dominant mode of propagation for MF radio waves in coastal areas and near shore, whereas the skywave is the dominant mode of propagation for MF radio waves in open ocean areas. HF radio waves in the 4 MHz through 26 MHz bands are characterized primarily by skywaves reflected by the ionosphere, which is the dominant mode of propagation for HF radio waves in all areas, both land and sea.

7.4.1 Digital Select Calling (DSC)

DSC capabilities available for maritime HF radios are essentially the same as those described in the previous paragraph for maritime VHF radio. DSC services, which may also be encrypted for security purposes, enable mariners to send or receive distress, urgency, safety and routine radiotelephone calls to or from any similarly equipped vessel or shore station, including automatic digitally formatted distress alerts to coast guards and rescue authorities anywhere in the world.

Operating the DSC feature is accomplished where the Maritime Mobile Service Identity (MMSI) and call sign programmed into the HF radio upon initial installation on the vessel are encoded within a digital message along with position information acquired using a built-in GNSS capability present within the radio itself, or obtained from an onboard GNSS received connected to the radio. The MMSI of the recipient vessel or shore station is entered into the radio keyboard and the message is initiated by pressing the "send" button.

7.4.2 Weather Broadcasts and Marine Safety Information

A variety of weather information and maritime safety information is available over MF and HF radio through NAVTEX, Radiofacsimile and radioteletype. High seas forecasts and storm warnings are also available from national sources worldwide from whom station locations and schedules may be obtained. Transmission range is dependent upon operating frequency,

Table 7.3 HF Single Sideband Maritime Radiotelephone Channels

	ITU Channel No.	*Coast Transmit*	*Ship Transmit*	*Notes*
4 MHz Duplex Channels	401–427	4.357–4.435 kHz	4.068–4.143 kHz	3 kHz spacing
4 MHz Duplex Channel	428	4.351 kHz	Varies	
4 MHz Duplex Channel	429	4.354 kHz	Varies	
4 MHz Simplex Channel	1 through 21	4.146–4.149 kHz 4.000–4.060 kHz		3 kHz spacing
4 MHz Shared Channels				3 kHz spacing
4 MHz (Calling)	**421**	**4.147 kHz**	**4.125 kHz**	**Distress and safety working 4.125 kHz simplex**
4 MHz (USCG Calling)	**424**	**4.435 kHz**	**4.134 kHz**	
6 MHz Duplex Channels	601–608	6.501–6.522 kHz	6.200–6.212 kHz	3 kHz spacing
6 MHz Simplex Channel		6.224–6.230 kHz		3 kHz spacing
6 MHz (Calling)	**606**	**6.215 kHz**		**Distress and safety working 6.225 kHz simplex**
8 MHz Duplex Channels	801–832	8.719–8.812 kHz	8.195–8.291 kHz	3 kHz spacing
Distress	**833**	**8.291 kHz**		**Distress and safety working 8.291 kHz simplex**
8 MHz Duplex Channels	834–837	8.707–8.716 kHz	Varies	
8 MHz Simplex Channel		8.294–8.297 kHz		3 kHz spacing
8 MHz (USCG Calling)	**816**	**8.764 kHz**	**8.240 kHz**	
8 MHz (Calling)	**821**	**8.779 kHz**	**8.255 kHz**	
12 MHz Duplex Channels	1201–1241	13.077–13.197 kHz	12.230–12.350 kHz	3 kHz spacing
12 MHz Simplex Channels		12.353–12.365 kHz		3 kHz spacing
12 MHz (USCG Calling)	**1205**	**13.089 kHz**	**12.242 kHz**	
12 MHz (Calling)	**1221**	**13.137 kHz**	**12.290 kHz**	**Distress working 12.290 kHz simplex**

(Continued)

Table 7.3 (Continued)

	ITU Channel No.	*Coast Transmit*	*Ship Transmit*	*Notes*
16 MHz Duplex Channels	1601–1656	17.242–17.407 kHz	16.360–16.525 kHz	3 kHz spacing
16 MHz Simplex Channels		16.528–16.546 kHz		3 kHz spacing
16 MHz (USCG Calling)	**1625**	**17.314 kHz**	**16.432 kHz**	
16 MHz (Calling)	**1621**	**17.302 kHz**	**16.420 kHz**	**Distress working 12.290 kHz simplex**
18/19 MHz Duplex Channels	1801–1815	19.755–19.797 kHz	18.780–18.822 kHz	3 kHz spacing
18/19 MHz Simplex Channels		18.825–18.843 kHz		3 kHz spacing
18/19 MHz (Calling)	**1806**	**19.770 kHz**	**18.795 kHz**	
22 MHz Duplex Channels	2201–2253	22.696–22.852 kHz	22.000–22.156 kHz	3 kHz spacing
22 MHz Simplex Channels		22.159–22.177 kHz		3 kHz spacing
22 MHz (Calling)	**2221**	**22.756 kHz**	**22.060 kHz**	
25/26 MHz Duplex Channels	2501–2510	26.145–26.172 kHz	25.070–25.097 kHz	3 kHz spacing
25/26 MHz Simplex Channels		25.100–25.118 kHz		3 kHz spacing
25/26 MHz (Calling)	**2510**	**26.172 kHz**	**25.097 kHz**	

time of day and atmospheric conditions and can vary from only short distances to several thousand miles.

The primary method of broadcasting Marine Safety Information (MSI) is NAVTEX, which can provide coverage up to around 400 nautical miles out to sea although this depends on many reception issues. All NAVTEX broadcasts are made on 518 kHz, using narrow-band direct printing 7-unit forward error correcting (FEC or Mode B) transmission. This service is complemented by radio telephony with networks of remote radio sites around coasts providing MF coverage of 150 miles.

HF radiofacsimile broadcasts originating from various worldwide locations are shown below.[16] All emissions are amplitude modulation, single sideband, suppressed carrier, analog (J3C) and frequencies are subject to change.

Australia:	2,628 kHz, 5,100 kHz, 5,755 kHz, 7,535 kHz, 10,555 kHz, 11,030 kHz, 13,920 kHz, 15,615 kHz, 20,469 kHz, 18,060 kHz.
Canada:	3,253.0 kHz, 4,271 kHz, 4,416 kHz, 4,292.0 kHz, 6,915.1 kHz, 6,496.4 kHz, 7,710.0 kHz, 8,456.0 kHz, 10,536 kHz, 13,510 kHz.
China:	4,199.75 kHz, 8,412.5 kHz, 12,629.25 kHz, 16,826.25 kHz.
Chile:	4,228.0 kHz, 4,322.0 kHz, 8,677.0 kHz, 8,696.0 kHz, 17,146.4 kHz.
Greece:	4,481 kHz, 8,105 kHz.
Germany:	3,855 kHz, 7,880 kHz, 13,882.5 kHz.
Japan:	148 kHz, 3,622.5 kHz, 7,795 kHz, 13,988.5 kHz.
Japan/Singapore:	4,316 kHz, 8,467.5 kHz, 12,745.5 kHz, 16,035 kHz, 16,971 kHz, 17,069.6 kHz, 17,430 kHz, 22,542 kHz.
Korea Republic:	3,585 kHz, 5,857.5 kHz, 7,433.5 kHz, 9,165 kHz, 13,570 kHz.
New Zealand:	3,247.4 kHz, 5,807 kHz, 9,459 kHz, 13,550.5 kHz, 16,340.1 kHz.
Russia:	4,481 kHz, 5,336 kHz, 6,446 kHz, 7,907 kHz, 7,908.8 kHz, 8,105 kHz, 8,444 kHz, 10,130 kHz.
Thailand:	7,395 kHz.
United Kingdom:	2,618.5 kHz, 4,610.0 kHz, 8,040.0 kHz, 11,086.5 kHz.
United States:	2,054 kHz, 4,298 kHz, 4,317.9 kHz, 4,346 kHz, 8,459 kHz, 8,503.9 kHz, 8,682 kHz, 12,412.5 kHz, 12,786 kHz, 12,789.9 kHz, 17,146.4 kHz, 17,151.2 kHz, 22,527 kHz.

7.4.3 Precision Time Broadcasts

Should GNSS time reference signals become unusable or are otherwise unavailable, there exist precision time signals broadcast worldwide that can be used for navigation purposes. A list of many of these signals and the frequencies on which they are broadcast appears below:

- Canada, CHU (Ottawa), 3.33, 7.85 and 14.67 MHz.[17]
- China, BPM (Pucheng), 2.5 MHz, 5 MHz, 10 MHz, 15 MHz.[18]
- Germany, DCF77 (Mainflingen), 77.5 kHz.[19]
- Japan, JJY (Tokyo), 40 kHz, 60 kHz.[20]
- South Korea, HLA (Daejeon), 5 MHz.[21]
- Taiwan, BSF (Taichung), 77.5 kHz.[22]
- United Kingdom, MSF (Cumbria), 60 kHz.[23]
- United States, WWV (Colorado Springs, CO), 2.5, 5, 10, 15 and 20 MHz.[24]
- United States, WWVH (Hawaii), 2.5, 5, 10, 15 and 20 MHz.[25]

7.4.4 HF Signal Propagation Effects

One of the greatest factors affecting radio signal propagation in the lower frequencies of the radio spectrum is the time of day and specifically, the effect of the sun's radiation on the ionosphere. High atmospheric ionization levels during the day can make it more difficult for radio waves to propagate, whereas at night the atmosphere is less ionized making it easier for radio waves to propagate. One great exception to this rule is the effect of sunspots combined with the usual daylight ionization effects that can inhibit normal local, short-distance communications while simultaneously making it possible to communicate with stations thousands of miles away. This results in "skip" conditions where radio waves repeatedly reflect between the surface and the ionosphere over great distances.

Seasons can also affect propagation with winter ionization levels generally higher and more difficult for radio waves to propagate than with lower summer levels. Weather can also affect propagation with high atmospheric pressure causing radio waves to propagate farther than during periods of low atmospheric pressure when radio waves, in general, propagate shorter distances. Geographic features can also affect propagation where mountains and other obstacles can block radio waves along with the curvature of the Earth, which can also limit radio wave propagation distance.

7.5 Global Maritime Distress and Safety System (GMDSS)

The GMDSS is an automated ship-to-shore and ship-to-ship distress and radio communication safety system that uses satellites and/or terrestrial radio systems with digital selective calling (DSC) technology.[26] Developed under the International Maritime Organization (IMO) and mandated by the SOLAS Convention for ships that are over 300 gross tons and also ships that carry more than 12 passengers internationally, it provides safety of life information and communication systems that inform vessels of navigation hazards and weather conditions, and enable distress calls with pertinent location and identification information with the push of a button. GMDSS is comprised of safety procedures, types of equipment and communication protocols used for safety and rescue operations of distressed ships, boats and aircraft. It also allows ships to be located quickly and efficiently in the event of a distress situation. A major component is the COSPAS-SARSAT satellite search and rescue (SAR) system, where instruments are flown on board LEO, MEO and GEO satellites provided by the United States, Russian Federation, India and the European Union.[27] These satellites are capable of detecting signals coming from the Earth's surface transmitted by an emergency Search and Rescue Transmitter (SART) that, when activated, broadcasts a distress beacon on a frequency of 406 MHz to send a signal that indicates the position of a distressed vessel.

All GMDSS ships must carry a 406 MHz Emergency Position Indicating Radiobeacon (EPIRB), a VHF radio capable of transmitting and receiving DSC and radiotelephony, a Navigational Telex (NAVTEX) receiver, a SART, backup power systems and two-way VHF portable radios. In GMDSS Sea Area 2, where radiotelephone coverage of at least one MF coast station in which continuous DSC (2,187.5 kHz) alerting and radiotelephony services are available, ships must also carry a DSC-equipped MF radiotelephone.[28] For all voyages conducted in GMDSS Sea Areas 3 and 4, vessels must carry either an Inmarsat F77, B or C ship earth station, or a DSC-equipped HF radiotelephone/telex.[28]

Companies such as Inmarsat/Viasat using radionavigation transponders can enable Satellite-Based Augmentation System (SBAS) services around the world for coast guard and other organizations to enhance standard GPS/Galileo accuracy of 5–10 meters to as little as 10 cm using satellite connectivity, land-based infrastructure and software.[30] Precise tracking such as this could enable pinpoint safety navigation and help emergency services reach vessels in distress more quickly.

7.6 Visual Communications

Several different means of visual communication are used to indicate ship function and activities to other vessels, to identify aids to navigation and to indicate distress situations at sea.

7.6.1 *Lights and Dayshapes*

A significant portion of the IMO Convention on the International Regulations for Preventing Collisions at Sea (COLREGS) Part C is dedicated to communication using visual signals to be complied with in all weathers (Rule 20). These signals have to do with the use of lights from sunset to sunrise, and from sunrise to sunset in restricted visibility and may be exhibited in all other circumstances when it is deemed necessary. Also specified is the use of dayshapes by day.

Knowledge of the content, meaning and implications of each of these signals is required of watchstanders and autonomous assistants on the bridge, and their use becomes secondhand during normal operations. An illustration of some of these lights and dayshapes is provided in Figure 7.1. All lights and dayshapes shown are international, and inland visual signals are likely to be different in accordance with the customs and practices of individual nations. Additional lights not shown but required when making way through the water include a masthead light or lights, sidelights and a stern light. Also not shown is an all-round flashing yellow light that can be found on an air-cushion vessel when operating in the non-displacement mode, and any special light or lights that may be used by a submarine on the surface.

a (Top) Pilot vessel – COLREGS Rule 29. (Middle) Sailing vessel underway, or single light with three quadrants: white, red and green – Rule 25. (Bottom) Vessel at Anchor – Rule 30, or power-driven vessel of less than 7 meters in length whose maximum speed does not exceed 7 knots – Rule 23.
b Vessel not under command or restricted in their ability to maneuver – Rule 27.
c Vessel constrained by her draft – Rule 28. Vessel aground – Rule 30.
d Vessel engaged in mine clearance operations (<50 m) – Rule 27.
e Vessel proceeding under sail when also being propelled by machinery, except that a vessel of less than 12 meters in length is not required to exhibit the dayshape – Rule 25.
f (Top) Vessel engaged in trawling. (Bottom) Vessel engaged in fishing other than trawling – Rule 26.
g Same as f. except additional daymark indicating gear extended beyond 150 meters horizontally from the vessel – Rule 26.

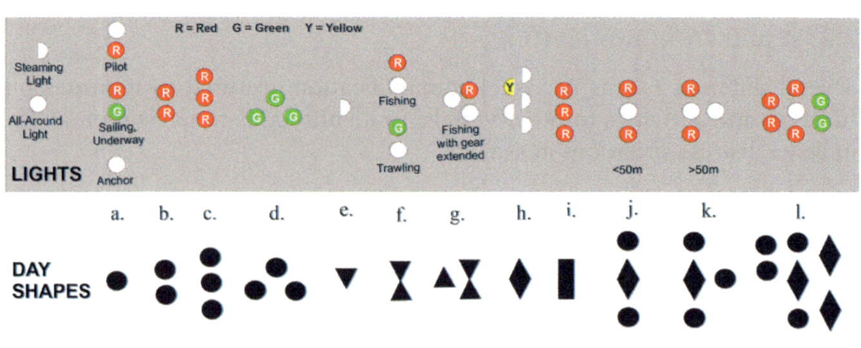

Figure 7.1 Masthead Lights and Dayshapes.

Notes: International Rules for the Prevention of Collisions at Sea, 1972 (COLREGS) as codified in the United States by Coast Guard Commandant Instruction M16672.2D Navigation Rules –International and Inland. Part C, Lights and Shapes. 25 March 1995.

h Power-driven vessel towing astern – towing vessel less than 50 meters in length; length of tow exceeds 200 meters – Rule 24.
i Vessel constrained by her draft – Rule 28.
j Vessel restricted in her ability to maneuver, or small vessel engaged in diving operations (<50 m) – Rule 27.
k Vessel restricted in her ability to maneuver, at anchor (<50 m) – Rule 27.
l Vessel engaged in dredging or underwater operations when restricted in ability to maneuver – not making way with an obstruction on the starboard (or port) side – Rule 27.

These specific visual signals when combined with navigation lights while making way that reflects various vessel configurations must be distinguished from deck lighting and other background lights likely to be encountered. This requires knowledge and ability that is relatively easily taught by educators and learned by mariners as part of their normal training and experience. However, in addition to the basic steering and sailing rules of the COLREGS Part B, these Lights and Shapes requirements of the COLREGS Part C must be included in all training regimens for crewed vessels and unmanned and autonomous vehicles using visible light cameras with the development process properly verified and the results validated and tested. This includes training encompassing all ship viewing perspectives at sea through 360° in varying visibility and weather conditions before such machines should be released onto the high seas and inland waterways. Some research in this regard has been published in the literature, but its scope is limited with further investigation still needed.[31]

7.6.2 Coded Light Signals from Aids to Navigation

Lighted code signals differentiated by color and rhythm are used to identify maritime aids to navigation, such as lighthouses, range lights, and buoys.[32] Lighthouses and range lights may display distinctive light characteristics to facilitate individual aid to navigation recognition that is verifiable using the Light List and navigation charts, but no special significance should be attached to the color or rhythm of such lights. Other lighted aids to navigation, such as buoys, employ light characteristics to convey additional information. The description below follows the International Association of Marine Aids to Navigation and Lighthouse Authorities (IALA) B international aid to navigation marking system used in the Americas and other parts of the world, and does not conform to the IALA A system used throughout Europe and many parts of Asia. However, the concepts regarding light signals are still applicable to communication regardless of whether the IALA A or B convention is used.

In much the same manner as for lights and shapes, lighted AtoN also display various characteristics essential to ensure proper identification and thereby position. These characteristics include green and red lights that mark the sides of channels and locations of wrecks or obstructions that are to be passed by keeping these lights on the port (left) hand of a vessel. Green and red lights are also used on Preferred Channel Marks, where the topmost band is green and red, respectively. Yellow or white lights having no lateral significance can also be displayed on AtoN, whose purpose may be determined by their shape, color, letters or numbers, and the light rhythm employed.

Light rhythms can vary greatly displaying both regularly flashing or regularly occulting light rhythms. Some examples include when flashing lights (frequency not exceeding 30 flashes per minute) may also be used, and for situations where lights require a distinct cautionary significance such as at sharp turns, sudden channel constrictions, wrecks, or obstructions, a quick flashing light rhythm (60 flashes or more per minute) may be used. Composite group flashing light rhythms consisting of groups of two flashes followed by one flash may be found on some AtoN in addition to Safe Water Marks that display a white Morse Code "A" rhythm (short-long flash) and Isolated Danger Marks that display a white group flashing two flashes. Other lights include those used on Special Marks that display yellow lights with fixed or slow flashing rhythms, mooring buoys and information and regulatory marks displaying white lights of various rhythms.

Lights containing sector guidance showing one color from most directions and a different color or colors over definite arcs of the horizon as indicated on the appropriate nautical chart are sometimes used to mark shoals or warn mariners of other dangers. Approximate bearing information is provided where the observer can note a change of color as the boundary between the sectors is crossed.

Some research in the area of computer-based recognition of AtoN lights and colors has appeared in the literature.[33] However, there still exists much work to be done before unmanned and autonomous vehicles will be capable of dealing with visual AtoN recognition at night. Further information on the IALA A and B system of buoyage and aids to navigation is available from IALA.[34]

7.6.3 International Code of Signals (ISC)

The International Code of Signals (ISC) considers situations related to the safety of navigation and persons, especially when language difficulties arise.[35] Useful for all modes of communication, it contains alphanumeric characters and visual flags representing specific information that may be combined with several flags, or semaphores, to send various messages. Methods of visual signaling messages may include the use of flags (see Figure 7.2),[34] flashing lights and sound signaling using Morse code symbols, voice over a loud hailer and Morse signaling by hand flags or arms. In addition, the use of radiotelegraphy

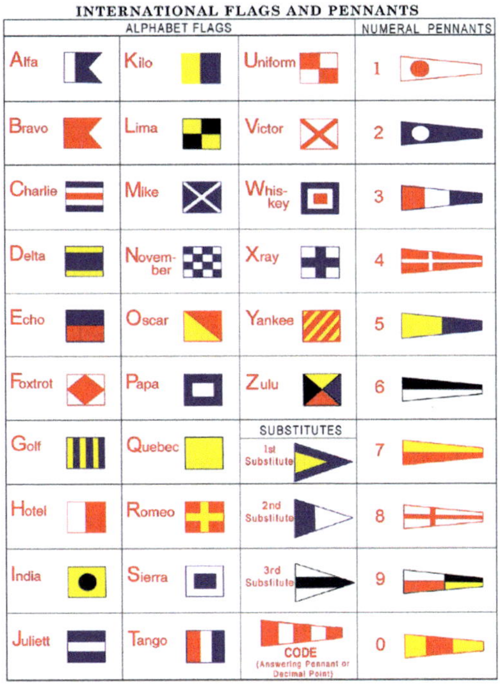

Figure 7.2 International Code of Signals, Flags and Pennants.

Source: Athul, M., Yogitha, A., and Nandakumar, R. Flag semaphore detection using Tensorflow and Opencv. International Journal of Recent Technology and Engineering (IJRTE), 7(6), ISSN: 2277-3878 (Online), March 2019.

and radiotelephony is appropriate for communicating characters and numbers to compose messages.

Signal flags are grouped according to alphabetic flags, numeric pennants, substitutes, repeaters and special flags using six prominent colors that are easily recognized including blue, yellow, red, black and white. The number of flags used to create a message also conveys meaning with one flag for urgent or common signals, two flags for distress and maneuvering, three flags for compass points, relative bearings, standard times, verbs and punctuation; four flags for geographical signals, names of ships, etc.; and five flags for time and position.

Examples of international meanings for signal flags include:[34]

A (Alpha)	Diver down.
B (Bravo)	Carrying dangerous cargo.
C (Charlie)	Yes (affirmative).

D (Delta) Keep clear of me. I am maneuvering with difficulty.
E (Echo) Altering course to starboard.
F (Foxtrot) I am disabled, communicate with me.
G (Golf) I want a pilot.
H (Hotel) A pilot on board.
I (India) I am altering my course to port.
J (Juliet) Vessel on fire keep clear.
K (Kilo) I want to communicate with you.
L (Lima) Stop your vessel instantly, I have something important to communicate.
M (Mike) My vessel is stopped and making no way through the water.
N (November) No (negative).
O (Oscar) Man overboard.
P (Papa) In port: All personnel return to ship; vessel is about to sail; At sea: It may be used by fishing vessels to mean: "My nets have come fast upon an obstruction".
Q (Quebec) I request free pratique.
R (Romeo) Reverse course.
S (Sierra) Engines are going astern.
T (Tango) Keep clear; engaged in trawling.
U (Uniform) You are heading into danger.
V (Victor) Require assistance.
W (Whiskey) Require medical assistance.
X (X-ray) Stop your intention.
Y (Yankee) Am dragging anchor.
Z (Zulu) I require a tug.

There are clearly hundreds of messages capable of being communicated using signal flags, and many mariners are proficient in their usage and meaning. Some research is also being performed to a limited extent regarding machine recognition and understanding of flags.[35,36] However, the scope of this research does not extend to knowledge and understanding of multiple flags communicating complex messages of the nature likely to be encountered by vessels at sea as would be useful for unmanned and autonomous vehicles.

7.6.4 *Visual and Audible Emergency and Distress Signals*

It is the duty of the master or individual in charge of a vessel to render assistance to any individual found at sea in danger of being lost, so far as the master or individual in charge can do so without serious danger to the master's or individual's vessel or to individuals on board. This implies that watchstanders and other crewmembers are trained in methods to communicate distress and in the understanding of their meanings.

Various means of attracting attention and securing assistance using visible and audible means have been devised should the need arise. These include a variety of signaling devices, both pyrotechnic and non-pyrotechnic, such as hand-held red flares and parachute flares for use both day and night and electric distress lights for the night. Day signals include hand-held and floating orange smoke signals, a signal mirror, water dye markers, a flag showing a square and ball displayed against an orange background, flying the NOVEMBER flag directly over the CHARLIE flag, and flying a square black flag with a black ball directly underneath. Additional visible signals can include the flashing of a high-intensity white light flashing at regular intervals from 50 to 70 times per minute, flames and/or smoke emanating from a vessel, a person on deck waiving their arms, the continuous sounding of a fog horn, the use of five prolonged blasts from a horn, signaling the Morse code "SOS" (• • • — — — • • •) from a horn (possibly supplemented using light signals in coordination with the horn), and the firing of a gun at regular intervals of one minute. Figure 7.3 illustrates several signals and methods that can be used to communicate distress situations at sea.

Figure 7.3 Visual Distress Signals.

Visual and audible signals within the line of sight of potential rescuers may provide the most expeditious response, but they may also be subject to misunderstanding by untrained observers and automated assistants that can result in a significant lag in time before the significance of distress signals are made clear, if at all. In addition to these local signals, distress signals should be sent to coast guards and regional search and rescue authorities using the more reliable and noticeable long-range emergency notification capabilities provided through VHF and HF radio communications, Inmarsat, and other satellite service providers.

Despite their effectiveness, the reliability of detection is reduced since monitoring the horizon to detect visual and audible distress signals is mainly accomplished by bridge watchstanders and observers who happen to gaze in the right direction at the right time. Smart visible light and infrared cameras with 360° coverage of the horizon and trained in flare, fire, and smoke detection using AI techniques can greatly assist in detecting such events through continuous monitoring and the ability to immediately notify watchstanders of such an occurrence. Camera coverage is becoming more widespread as onboard security interests increase, and integration of visual signal comprehension technology can be accomplished relatively easily. Likewise, monitoring of audio to detect and locate gunfire is already possible and in use throughout many of the larger cities of the world. Similar methods can be used to train smart microphones and audio systems to detect ship horns emitting warning, danger and distress signals.

7.7 Sound and Light Signals

In much the same manner as for visual and audible distress signals, Part D of the COLREGS addresses sound and light signals and states that devices installed on ships capable of producing loud sounds can include a horn (whistle), bell, and gong. Such sounds, possibly in conjunction with light signals, can coordinate passage between two vessels at close quarters, warn of danger, or reveal own vessel's presence and position under reduced visibility conditions. Sounds must be produced using any sound signaling appliance capable of producing the a "short blast" of about 1 second's duration, or a "prolonged blast" lasting from 4 to 6 seconds' duration. In addition to human sight and sound, high-resolution cameras and sensitive directional microphones can supplement human capabilities in detecting and understanding these signals. Methods of using these sounds are described in the following paragraphs.

7.7.1 Maneuvering and Warning Signals

When vessels are in sight of one another, a power-driven vessel underway shall when maneuvering, indicate the nature of the maneuver by the following signals (and, if equipped, may be supplemented by an all-round white visible at a minimum range of 5 miles while the maneuver is being carried out):[37]

- One short blast (and light flash of at least one second's duration, if equipped) to mean, "I am altering my course to starboard".
- Two short blasts (and light flashes, if equipped) to mean, "I am altering my course to port".
- Three short blasts (and light flashes, if equipped) to mean, "I am operating astern propulsion".

A yellow light may be used in lieu of a white light in some world regions depending on national regulations.

When vessels are in sight of one another in a narrow channel or fairway, a vessel intending to overtake another shall indicate intentions by the following signals:

- Two prolonged blasts followed by one short blast (and light flashes, if equipped) to mean, "I intend to overtake you on your starboard side".
- Two prolonged blasts followed by two short blasts (and light flashes, if equipped) to mean, "I intend to overtake you on your port side".

The vessel about to be overtaken shall indicate her agreement by the following signal on her whistle:

- One prolonged, one short, one prolonged and one short blast (and light flashes, if equipped), in that order.

When vessels are in sight of one another and approaching each other where, from any cause, either vessel fails to understand the intentions or actions of the other, or is in doubt whether sufficient action is being taken by the other to avoid collision, the vessel in doubt shall immediately indicate such doubt by giving at least five short and rapid blasts (and light flashes, if equipped).

Also, a vessel, when nearing a bend or an area of a channel or fairway where other vessels may be obscured by an intervening obstruction shall sound one prolonged blast. Such signal shall be answered with a prolonged blast by any approaching vessel that may be within hearing around the bend or behind the intervening obstruction.

These signals presuppose proper training of seafarers in the detection of horns and lights that includes the ability to discern their presence in the midst of background clutter noise. Bridge aids for watchstanders along with viewing and listening technology on unmanned and autonomous vehicles using directional microphones can narrow the relative bearing of the signal, and one or more cameras can be used to positively identify the signaling vessel. Training of AI-based smart microphones using the specific horn blast patterns and frequencies designated for use by horns, bells and gongs combined with smart camera training using these same patterns represented by white lights can today make possible the automated detection and identification of these signals using existing technology.

7.7.2 Sound Signals in Restricted Visibility

Special sounds are designated for use by vessels in areas and during times of restricted visibility during both day and night conditions.[38] Under such conditions in or near an area of restricted visibility, the following signals are mandated, as appropriate:

- A power-driven vessel making way shall sound at intervals of not more than 2 minutes one prolonged blast.
- A power-driven vessel underway but stopped and not making way shall sound at intervals of not more than 2 minutes, two prolonged blasts in succession with an interval of about 2 seconds between them.
- A vessel not under command, a vessel restricted in her ability to maneuver, a vessel constrained by her draft, a sailing vessel, a vessel engaged in fishing, and a vessel engaged in towing or pushing another vessel shall sound at intervals of not more than 2 minutes three blasts in succession consisting of one prolonged followed by two short blasts.
- A vessel engaged in fishing, when at anchor, and a vessel restricted in her ability to maneuver when carrying out her work at anchor, shall instead sound at intervals of not more than 2 minutes three blasts in succession consisting of one prolonged followed by two short blasts.
- A vessel towed or, if more than one vessel is towed, the last vessel of the tow, if manned, shall at intervals of not more than 2 minutes sound four blasts in succession, namely one prolonged followed by three short blasts. When practicable, this signal shall be made immediately after the signal made by the towing vessel.
- When a pushing vessel and a vessel being pushed ahead are rigidly connected in a composite unit they shall be regarded as a power-driven vessel and shall give the signals appropriate for being underway, and making way and not making way.
- Additional signals using bells and gongs are also identified for use by vessels when at anchor, aground and pilot vessels performing pilotage duties.

Although in or near areas of restricted visibility the use of lights may be of limited utility, audio monitoring using sensitive directional microphones can greatly improve situational awareness that can complement Radar and

AIS sensor data. Training of smart microphones on the specific patterns and frequencies mandated for horn, bell and gong use under these circumstances can result in speedy detection and positive identification of nearby vessels.

7.7.3 Signals to Attract Attention

There are times when it may be necessary to attract the attention of another vessel.[39] This should be accomplished using light or sound signals that cannot be mistaken for any authorized signal, or by directing a searchlight beam in the direction of the danger, in such a way as not to embarrass any vessel. Any such signal must not be mistaken for any aid to navigation, and the use of high-intensity intermittent or revolving lights, such as strobe lights, shall be avoided.

Automated computer-based detection and understanding of visual and audible signals intended to attract attention would follow the same general processes as for other light and sound signals. However, the range of improvisation and variability of such signals may make this task difficult.

7.8 Summary

This chapter has presented many different modes of communication as well as visual and audible messaging and content that can take years for trained mariners to master. Research has been performed but few case studies exist where attempts to automate individual tasks have been overwhelmingly successful, and such efforts continue. However, these technologies are at best still experimental and as of yet are unproven.

Also notable is the fact that, with the introduction of unmanned and autonomous technologies into the maritime mix, it appears that exceptions are being made at the national levels to eliminate the need to possess full knowledge of the requirements, capabilities, applications and means of use for all communication methods expected of seagoing vessels. The full significance in granting such exceptions to overall maritime safety and safety of navigation is yet to be determined.

References

1 Gavin Wright. Metropolitan Area Network (MAN), definition. TechTarget. February 2021. www.techtarget.com/searchnetworking/definition/metropolitan-area-network-MAN
2 Mike Wall, 2022. SpaceX Gets Permission to Deploy 7,500 Next-Generation Starlink Satellites. Space.com. 2 December 2022. www.space.com/spacex-fcc-approval-7500-starlink-satellites
3 Mike Schuler, 2023. Viasat Completes Acquisition of Inmarsat. gCaptain. 31 May 2023. https://gcaptain.com/viasat-completes-acquisition-of-inmarsat/

4 Marcus Hand. Are Hybrid VSAT and LEO Networks the Next Maritime Communications Solution? *Seatrade Maritime News*. 02 June 2023. www.seatrade-maritime.com/technology/are-hybrid-vsat-and-leo-networks-next-maritime-communications-solution?

5 US Coast Guard, 2008. Coast Guard Announces Successful Launch of Nationwide AIS Satellite. *Coast Guard News*. 23 January 2002, https://coastguardnews.com/coast-guard-announces-successful-launch-of-nationwide-ais-satellite/

6 Daniel O'Donohue. How Cargo Ships and Marine Vessels Are Tracked. Mapscaping. 9 September 2022. https://mapscaping.com/tracking-ships-and-vessels/

7 Rolls-Royce, 2017. Rolls Royce Demonstrates Worlds First Remotely Operated Commercial Vessel. Press Release. 20 June 2017. www.rolls-royce.com/media/press-releases/2017/20-06-2017-rr-demonstrates-worlds-first-remotely-operated-commercial-vessel.aspx

8 Wärtsilä, 2017. Successful Test of Remote Control Ship Operating Capability. Wärtsilä Marine Solutions. Wärtsilä Electrical and Automation. September 2017. www.wartsila.com/docs/default-source/marine-documents/business-white-papers/white-paper-o-ea-remote-control-ship.pdf?sfvrsn=16669e45_6

9 Mayflower 2022. It's Time for the Mayflower Autonomous Ship. www.mas400.com/technology

10 Iridium, 2021. The Mayflower Autonomous Ship Transforms Ocean Science with the Help of Iridium. Iridium Communications Inc, Press release, 27 May 2021. https://investor.iridium.com/2021-05-27-The-Mayflower-Autonomous-Ship-Transforms-Ocean-Science-with-the-Help-of-Iridium-R

11 US Coast Guard, 2023. VHF International Transmitting Frequencies. US Coast Guard Navigation Center. Department of Homeland Security. www.navcen.uscg.gov/international-vhf-marine-radio-channels-freq

12 DSC, 2023. United States Coast Guard, Navigation Center. U.S. Department of Homeland Security. Digital Select Calling. https://navcen.uscg.gov/digital-selective-calling

13 Rescue 21. U.S. Coast Guard, Acquisitions Directorate. Homeland Security. www.dcms.uscg.mil/Our-Organization/Assistant-Commandant-for-Acquisitions-CG-9/Programs/C4ISR-Programs/Rescue-21/

14 EMSA, 2023. Enhanced Maritime Picture via Integrated Maritime Services. European Maritime Safety Agency. www.emsa.europa.eu/we-do/digitalisation/maritime-monitoring.html

15 Electronics Desk, 2023. Duct Propagation. https://electronicsdesk.com/duct-propagation.html

16 Worldwide Radiofacsimile Broadcast Schedules, National Oceanographic and Atmospheric Administration (NOAA), Bethesda, MD, USA. 21 October 2022.

17 CHU. National Research Council. https://nrc.canada.ca/en/certifications-evaluations-standards/canadas-official-time/nrc-shortwave-station-broadcasts-chu

18 BPM. Chinese Academy of Sciences. www.sigidwiki.com/wiki/BPM

19 DCF. Physikalisch-Technische Bundesanstalt (PTB), www.ptb.de/cms/en/ptb/fachabteilungen/abt4/fb-44/ag-442/dissemination-of-legal-time/dcf77/

20 JJY. National Institute of Information and Communications Technology. https://jjy.nict.go.jp/news/index-e.html

21 HJA. Korea Research Institute of Standards and Science. https://en.wikipedia.org/wiki/HLA_(radio_station)

22 BSF. National Time and Frequency Standards Laboratory. www.wikiwand.com/en/BSF_(time_service)

23 MSF. National Physical Laboratory (NPL). www.npl.co.uk/msf-signal

24 WWV. National Institute of Standards and Technology (NIST). www.nist.gov/pml/time-and-frequency-division/time-distribution/radio-station-wwv

25 WWVH. National Institute of Standards and Technology (NIST). www.nist.gov/pml/time-and-frequency-division/time-distribution/radio-station-wwvh

26 GMDSS. Global Maritime Distress and Safety System (GMDSS). U.S. Federal Communications Commission. www.fcc.gov/wireless/bureau-divisions/mobility-division/maritime-mobile/ship-radio-stations/global-maritime

27 Cospas-Sarsat System Overview. National Oceanographic and Atmospheric Administration. Bethesda, MD, USA. www.sarsat.noaa.gov/cospas-sarsat-system-overview/

28 US Coast Guard, 2023a. GMDSS Areas and Search and Rescue Areas. US Coast Guard Navigation Center. Department of Homeland Security. https://navcen.uscg.gov/gmdss-areas-and-search-and-rescue

29 gCaptain, 2023. Inmarsat to Launch I-8 Satellites to Power L-Band Network. gCaptain Daily. 19 May 2023. https://gcaptain.com/inmarsat-to-launch-i-8-satellites-to-power-l-band-network

30 Xin Hu, Yingjun Zhang, and Shaobo Wang. Ship Lights Recognition Using Image Processing. 20th COTA International Conference of Transportation Professionals. August 2020. DOI: 10.1061/9780784482933.039

31 USC Part 62. United States Code, Part 62, Subpart B. U.S. Aids to Navigation System. Paragraph 62.45, Light Characteristics.

32 Xu Han, et. al. Multilabel Video Classification Model of Navigation Mark's Lights Based on Deep Learning Computational Intelligence and Neuroscience. Volume 2021, Article ID 6794202, 13 pages. https://doi.org/10.1155/2021/6794202

33 IALA Pubs. AtoN Design and Delivery. Stnd. 1020. www.iala-aism.org/content/uploads/2021/02/IALA-Technical-documents-Catalogue-Edition2-final-compressed-more-min.pdf

34 Pub 102. International Code of Signals for Visual, Sound and Radio Communications. United States Edition. Pub 102. 1969 Edition (revised 2020). National Geospatial-Intelligence Agency.

35 Athul Motty, Yogitha A and R Nandakumar. Flag semaphore detection using Tensorflow and Opencv. International Journal of Recent Technology and Engineering (IJRTE), 7(6), ISSN: 2277-3878 (Online), March 2019.

36 Ming Gu, Kun Hao, and Zhiyi Qu. Flag detection with convolutional network. CSAI'18: Proceedings of the 2018 2nd International Conference on Computer Science and Artificial Intelligence, December 2018, pp. 258–262. https://doi.org/10.1145/3297156.3297159

37 COLREGS, Rule 34. IMO Convention on the International Regulations for Preventing Collisions at Sea, 1972. Rule 34. Maneuvering and Warning Signals.

38 COLREGS, Rule 35. IMO Convention on the International Regulations for Preventing Collisions at Sea, 1972. Rule 35. Sound Signals in Restricted Visibility.

39 COLREGS, Rule 36. IMO Convention on the International Regulations for Preventing Collisions at Sea, 1972. Rule 36. Sound Signals to Attract Attention.

Chapter 8

Intraship and Internal Communications

The previous chapter discussed how sensor and other data may be communicated beyond the confines of the vessel itself to other ships and shore facilities using satellite, radio, visual and other methods. However, internal communication of sensor data representing functions associated with engineering, navigation, cargo, and human activity comprise a very large subset of overall vessel communications. Security and crowd control technologies found worldwide at military installations, financial institutions, government facilities and even amusement parks have also been tailored adapted to these same applications on ships. This is accomplished using a combination of dedicated circuits and common data bus architectures utilizing both wired and wireless transmission modes tailored to the specific needs of the sensors communicating their measurements and observations.

The independent and self-contained environment found onboard surface ships and vehicles provides a unique combination of data communication requirements not found elsewhere. These requirements are supported using protocols, including Ethernet, developed to support high-speed data communications between computers and peripheral devices such as disk drives and printers. These protocols are general in nature and have been adapted to accommodate the needs of sensors and sensor systems. Others, such as the Controller Area Network (CAN) and their derivatives that include the National Marine Electronics Association (NMEA) 0183 and 2000 data bus architectures, are tailored to support the maritime-specific needs of engineering and navigation sensors as they evolved to accommodate expanding requirements associated with vessel automation and digitalization.

Data bus technology developed for the operation and automation of industrial control systems that include RS-485 form the basis for monitoring many different types of bulk, liquid, gas and other types of cargo found onboard vessels of all sizes, types and kinds. USB (Universal Serial Bus) and Bluetooth wireless technology provide access to individual network system devices that can be a gateway to the entire ship's network infrastructure. Wireless Fidelity (Wi-Fi) technology enables wireless access to the Internet and local network resources by enabling devices to connect to a local area network (LAN).

DOI: 10.1201/9781003377900-8

This chapter hopes to bring together some of these concepts in a logical way that will convey the unique requirements of ship sensors and how their needs are considered in the performance of their everyday functions. This is accomplished by describing some essential factors that must be considered when configuring sensor communications using bus architectures, followed by a high-level comparison between the different standards and protocols used in shipboard and surface vehicle environments. Details are then provided describing the most prevalent data bus architectures used throughout the maritime industry.

8.1 Integrated Bridge Systems

An integral part of any discussion on internal ship communications must begin with the ultimate goal, which is the existence of an integrated bridge system (IBS). This is defined as a combination of interconnected systems that allow centralized access to sensor information or command/control from workstations, with the aim of increasing safe and efficient ship management by suitably qualified personnel.[1] They are designed to perform two or more operations associated with passage execution, communications, machinery control; loading, discharging and cargo control; and safety and security. There is a commonality in the navigation-related information needed for safe passage along a route presented by an Integrated Navigation System (INS) and an IBS, often on shared or common workstations. However, an IBS considers an even broader range of engineering and navigation sensor data to fulfill ship management needs by qualified personnel in the overall safe and efficient operation of the vessel.

As a rule, sensor technology development and the range of their applications have always long outpaced instruments developed by IMO that address their use. IBS is not an exception to this rule. SOLAS chapter V, adopted in December 2000, cites in Regulation 19 carriage requirements that limit IBS scope to shipborne navigational systems and equipment under the purview of the officer in charge of the navigational watch.[2] This chapter addresses those engineering and navigation sensors directly integrated into IBS. In the near term, passenger and cargo sensors are not likely to be considered to any great extent in the design of these systems. This does not mitigate the fact that many of these sensors report data on factors that can have significant implications in terms of passage execution, safety and security that must not be ignored.

8.2 Essential Factors in Sensor Data Communications

The needs of sensor data communications overlap and are shared across many of the same systems, with their overall efficiency being highly dependent upon several factors that include:

- Data Quality
- Data Rate
- Latency
- Bandwidth
- Security
- Flexibility

In addition, the physical characteristics of the data bus can also affect its performance. For example, the length of the data bus and the distance data must travel can affect latency, and the type of cable used can affect bandwidth characteristics. Specific factors most important for a particular application depend on the inherent requirements associated with the sensors attached to the bus, the nature of sensor data communicated therein, the length of the communications channel and other aspects of the application. As a rule, a sensor data bus featuring a low error rate, high data rate, low latency, high bandwidth and strong security will be more powerful than a data bus with lesser performance characteristics. Note that these factors are associated with the communication channel itself and are considered independent and apart from sensor performance characteristics identified in Chapter 2 (2.3) including linearity, range, accuracy, range, sensitivity, etc.

8.2.1 Data Quality

The quality of sensor data is based in part upon the frequency of occurrence and the severity of errors that exist within the data. Error in terms of sensor measurements can be defined as "the result of a measurement minus the true value of the measurand".[3] Several types of errors are related to sensor data quality, including outliers, missing data, bias, drift, noise, constant value and uncertainty conditions.[4]

Outliers, or anomalies associated with measurement values or data points that differ significantly from other observations, are one of the greatest errors in terms of frequency of occurrence and significance. These can be caused by failures in proper measurement by the instrument itself through improper scaling. They can also be due to the effects of the environment on the data bus used as the communication channel caused by temperature, humidity, interference, poor connectivity or other influences. Another source of outliers can include random errors caused by arbitrary fluctuations in the data bus. For example, the measurement may be affected by vibrations or electrical noise. Outliers may also be created by novel data or data points of a different distribution than the rest of the data set due to a momentary interruption of data or uncharacteristic data. In all cases, outliers can have a significant impact on statistical data analytics that can lead to inaccurate results.

Missing data or data that is not otherwise available for an individual observation or series of observations is likely to be attributed to sensor malfunction, breakdown in the communication channel, or environmental factors where the data collected may have been too extreme causing it to be lost or corrupted. Missing data can have a significant impact on sensor data analytics, where the absence of a large number of data points may cause errors leading to inaccurate conclusions.

Measurement bias refers to systematic error whose outcome includes measurement results that are consistently too high or too low. This can occur due to errors within the sensor itself due to improper calibration and environmental factors affecting the data bus that can skew measurements and affect their accuracy. Multiple methods to acquire sensor data, as well as the use of multiple sensors, can help to reduce measurement bias. Implementing more effective means to minimize the effects of environmental factors can also help in reducing measurement bias.

Drift that can cause a gradual and cumulative change in the output of a sensor over time, thereby corrupting the signal, can be introduced into a sensor data bus through noise and other errors in the measured signal. Sensor data bus malfunction can also introduce errors due to electrical interference, timing errors, or other factors causing the sensor to become unstable and can lead to drift over time.

The sensor data bus can be a source of electrical noise that can interfere with sensor measurements and corrupt sensor data. This noise can be caused by the high-speed switching of transistors or other electrical interference by improperly shielded data bus controller hardware.

Errors caused by electrical noise, timing problems, drift, noise and loss of data bus stability can lead to reduced confidence in the degree to which the measurement can be trusted to be accurate. Uncertainty can be quantified in terms of a confidence level (in percent) wherein measurement values are likely to contain the true value of the measurement. Uncertainty can also be illustrated by the standard deviation in the spread of measurement values in a data set, where a close spread would indicate greater certainty in the measurement as opposed to a widespread that would indicate greater uncertainty.

8.2.2 Data Rate

Computer system data rate refers to the speed at which data is transferred within the computer or between a peripheral device and the computer, and is measured in different scales of bytes per second.[5] For sensor applications, data rate is the speed at which measurement data is sent across a communication channel. One of the key considerations includes the type of sensor and the nature of the data, which can vary significantly between different types of sensors. For example, a Global Navigation Satellite System (GNSS) or Inertial Navigation System (INS) may produce new data at a millisecond rate. At the same time, a temperature or level sensor may update its data over periods measured in seconds. Also, imagery data produced by a camera is often updated 20 or more times per second, with each update producing large amounts of data that must be communicated at a high speed to maintain flow and continuity to the viewer or real-time image analysis processors when capturing fast-changing phenomena. Each individual sensor has specified resolution and accuracy requirements that must be supported by available bandwidth and network capacity to avoid data congestion and potential data loss. Another aspect of this issue is the capability of the network to store and process the incoming data, where higher data rates may require larger storage capacities and efficient data management strategies. These issues must also be considered within the framework of existing regulatory requirements and standards that dictate specific data rates to ensure interoperability exists between instruments of different manufacturers.

8.2.3 Bandwidth

Whereas data rate is the speed at which a sensor produces data to be sent across a communication channel, bandwidth is the maximum amount of data that can be transferred over the data bus at a given time. Bandwidth is generally measured in terms of kilobits per second (kbps), megabits per second (Mbps) or gigabits per second (Gbps). A higher bandwidth allows more data to be transferred simultaneously, which can improve system scalability. It is critical that data bus bandwidth is adequate to support substantially more than the maximum amount of data that may be produced by the sensors integrated into the system to ensure the system does not get overwhelmed with excessive data that will bog down sensor system performance.

8.2.4 Latency

Latency is the delay in time between when sensor data is generated and when it is delivered to the end user, be it a human being or another electronic system whose purpose is to process the data. Lower latency allows data to be transferred more quickly, which can improve the responsiveness of the system. Adequate bandwidth capacity to handle anticipated sensor data rate requirements is necessary to ensure network latency is minimized.

8.2.5 Security

Data bus security is paramount to protect sensor data from unauthorized access and to prevent sensor system hacking and hijacking. A secure data bus uses encryption, authentication, access control, firewalls, intrusion detection and prevention systems, secure protocols, and other security measures to protect the data and the sensors that generate the data. Robust security measures require a comprehensive approach that considers all aspects of the sensor data bus application, from physical security to network protocols and access controls. Regular updates and improvements should be made to stay ahead of evolving security threats.

8.2.6 Flexibility

The flexibility of a data bus is a measure of how it may be adapted to different applications and configured to meet the specific requirements of the application. However, unique data bus architectures have been designed to optimize specific types of sensor data. For example, the CAN bus formed the basis for the creation of the NMEA-2000 bus, which is designed specifically for maritime use by electronics associated with navigation and engineering sensors. A similar adaptation known as J1939 was also created by the Society of Automotive Engineers (SAE) for commercial vehicle use.[6] CAN proved to be quite flexible in its design but didn't meet the specific needs of either the maritime or automotive communities and was adapted to each application.

8.3 LAN (Local Area Network)

A LAN is a group of computers and other devices that are connected together within a limited area, such as a ship, and allow devices to communicate with each other. LANs are also used to share resources such as printers, files and Internet access, as well as to connect sensors. Several different protocols can be used for connecting devices on a LAN including Ethernet, Bluetooth, and Wi-Fi.

LANs can be used to connect sensors on a ship in a number of ways. One way is to use a dedicated LAN that is specifically for sensors. This LAN would be isolated from other networks on the ship, which would help to protect the sensors from unauthorized access. Another way to connect sensors to a LAN is to use a wireless network. This would allow the sensors to be placed anywhere on the ship without the need for cables. However, wireless networks are more susceptible to interference than wired networks, so it is important to choose a wireless protocol that is well-suited for the environment.

8.4 Overview of Sensor Data Bus Standards and Protocols

There are several specific standards that apply to sensor connectivity on ships to ensure reliable and safe communication between sensors and the ship's systems. The choice of sensor data bus designation depends on the specific requirements of each application. It is important to note that sensor connectivity on ships can vary depending on the type of vessel, its systems and the specific sensors being used. Some of the key standards include:

- Ethernet
- CAN

- NMEA-0183
- NMEA-2000
- RS-485
- USB Port
- Bluetooth
- Wi-Fi

Table 8.1 provides a rough side-by-side comparison between the several bus and protocol standards and their characteristics described in this chapter in an attempt to highlight the contrast between them. However, comparing standards with different capabilities can lead to misleading conclusions and oversimplification. The intended goal is to highlight their use in the shipboard maritime environment and not to paint a complete picture of each standard's overall strengths and weaknesses. Direct comparisons are difficult as each standard serves different purposes and operates in diverse environments. Note also that this is only a representative sample of applicable standards, and other relevant standards have been left out of this comparison.

Ethernet is a widely used technology that defines the rules and standards for transmitting data over a LAN. This includes wired and fiber optic networks that support a wide range of bandwidth and data rates.

The CAN network is a type of communication protocol and network architecture commonly used in many applications including passenger and commercial vehicles, and in the maritime and industrial sectors. It was developed to address the need for a robust and reliable communication system for electronic control units (ECUs) in vehicles.

NMEA-0183 is a maritime-specific standard communication protocol developed by the NMEA for transmitting and receiving data between marine electronic devices. NMEA 0183 has been widely adopted in the marine industry. It is used for interconnecting navigation equipment, sensors and other marine electronics on recreational boats, ships of all kinds and sizes, and a wide variety of other watercraft.

NMEA-2000 evolved from the NMEA-0183 standard to provide significant improvements in terms of data speed, capacity and network capabilities to allow marine electronics, sensors and systems to communicate and share data over a single network.

RS-485 is an industrial specification that defines the electrical interface and physical layer for point-to-point communication of electrical devices and industrial sensors, many of which are used on ships. It specifies the electrical characteristics of a balanced differential serial communication interface, which allows for multi-point communication and long-distance data transmission.

The Universal Serial Bus (USB) standard is a widely used communication protocol and physical interface for connecting various devices to a computer

Table 8.1 Comparison between Shipboard Sensor Data Bus Standards

Characteristic	*Standard*									
	Ethernet			*CAN*	*NMEA 0183*	*NMEA 2000*	*RS-485*	*USB 3.1*	*Bluetooth 5.x*	*Wi-Fi*
	100BaseTX	*1000BaseT*	*1GBaseT*							
Data Rate	100 Mbps	1000 Mbps	10 Gbps	1 Mbps	9.8 kbps	250 kbps	up to 10 Mbps	10 Gbps	2 Mbps	11–3000 Mbps
Data Format	Binary	Binary	Binary	Binary	ASCII	Binary	Various	Serial packet	Packet	Binary
Protocol	IEEE 802.3	IEEE 802.3	IEEE 802.3	Message	Sentence	Message	Various	Master/ Slave	Master/ Slave	IEEE 802.11
Compatibility	Widespread	Widespread	Widespread	Moderate	Moderate	Widespread	Widespread	Widespread	Widespread	Widespread
Topology	Several	Several	Several	Bus	Point-to-Point	Star	Several	Low	Star	Star
Fault Tolerance	Moderate	Moderate	Moderate	High	Low	Low	High	Moderate	Moderate	Moderate
Security	High	High	High	Moderate	Low	High	None	Low	Moderate	Moderate

or other host devices.[7] On board ships it is generally used for updating navigation charts and maintenance manuals, and for servicing many types of electronic equipment.

Bluetooth is a wireless communication technology that allows electronic devices to exchange data over short distances without the need for physical connections. It is used in much the same manner as USB, except that it does not have to be physically connected for access to be achieved to a Bluetooth device.

Wireless Fidelity (Wi-Fi) is a wireless communication technology that allows devices to connect to a LAN and the Internet without the need for wired connections. It is based on the IEEE 802.11 family of standards, which defines the specifications for wireless networking.

Many standards also fall under the purview of the International Electrotechnical Commission (IEC), which develops and publishes international standards for electrical, electronic and related technologies, including:

- IEC 60945: Standards for regulating all equipment used in marine areas.
- IEC 61158: A series of related standards that cover aspects of RS-485 that define the fieldbus standard for digital communication in industrial automation.
- IEC 61162 (Maritime Navigation and Radiocommunication Equipment and Systems): A series of standards that specify the data communication protocols used for shipboard sensors and systems.
- IEC 61162-1: General requirements for data acquisition systems for marine applications covering the physical layer, the data link layer and the application layer for NMEA-0183.
- IEC 61162-2: Requirements for data acquisition systems for marine applications that use the Ethernet protocol.
- IEC 61162-3: Requirements for data acquisition systems for marine applications that use the CAN bus protocol and the NMEA-2000 implementation of this standard.

Details on these individual standards are provided in the following paragraphs.

8.5 Ethernet

Ethernet is a family of wired computer networking technology and protocols commonly used on ships for connecting devices on a LAN. First introduced in 1980 and standardized in 1983 as IEEE 802.3 and now also referred to as ISO/IEC8802-3, it uses physical media that includes twisted-pair copper wiring, coaxial cable and fiber optic cable. The data rate of Ethernet varies depending on the type of physical media and the version of the standard.

Older versions of the standard that are no longer used in modern networks include 10Base-T and 10Base-F, which are limited to speeds up to 10 Mbps. Relevant contemporary standards include:[8]

- 100BaseTX and 100BaseT4 (Cat 3 or higher unshielded twisted-pair copper wiring; 100 Mbps),
- 100BaseFX (fiber optic cable; 100 Mbps),
- 1000BaseT (Cat 5 or higher unshielded twisted-pair copper wiring; 1,000 Mbps) and
- 10GBaseT (Cat 5 or higher unshielded twisted-pair copper wiring; 10,000 Mbps or 10 Gbps).

Ethernet hubs were used to connect multiple Ethernet devices in a LAN, but these have been largely replaced by Ethernet switches due to their superior performance and features. Ethernet hubs operate at the physical layer (Layer 1) of seven layers described in the Open Systems Interconnection (OSI) model computer systems use to communicate over a network.[9] Ethernet switches operate at the physical layer (Layer 1) and the data link layer (Layer 2) to provide intelligent data forwarding and dedicated bandwidth to each connected device, making them far more efficient and scalable than older Ethernet hubs.

Also used are routers used to connect multiple networks and direct data traffic between them. A router operates at the network layer (Layer 3) of the OSI model and plays a critical role in facilitating communication between devices and networks in complex network environments. Some Ethernet switches offer Power over Ethernet (PoE) capabilities, providing power to connected devices and sensors such as IP cameras and wireless access points, eliminating the need for separate power supplies.

The OSI and Transmission Control Protocol/Internet Protocols (TCP/IP) are two different but related networking models associated with the Ethernet protocol. The OSI model is a conceptual framework that standardizes how different networking protocols should interact and communicate with each other, while TCP/IP is a suite of specific set of protocols used to interconnect network devices on the Internet.[10] While serving different purposes, the OSI and TCP/IP models have not entirely replaced the other. TCP/IP has become the dominant suite of protocols used for Internet communication, while the OSI model remains a basic reference model for understanding networking concepts and protocol interactions.

8.6 Controller Area Network (CAN) Bus

The CAN sensor data bus is a type of serial communication bus used to transmit data between ECUs on a ship. The CAN data bus follows a serial communication protocol to transmit data between ECUs in vehicles and other control applications such as ships. CAN is a serial communication protocol that supports distributed real-time control and multiplexing for use within vehicles and other applications. The standard defines the data format and protocol that are used to transmit data over the bus and specifies the Classical CAN frame format and the newly introduced CAN Flexible Data Rate Frame format. The Classical CAN frame format allows bit rates up to 1 Mbit/s and payloads up to 8 bytes per frame. The Flexible Data Rate frame format allows bit rates higher than 1 Mbit/s and payloads longer than 8 bytes per frame. Also described is the general architecture of CAN in terms of hierarchical layers according to the ISO reference model for open systems interconnection (OSI) according to ISO/IEC 7498-1.[11]

In the restructured ISO 11898 series most applicable to maritime applications:

- Part 1 defines the data link layer, including the logical link control (LLC) sub-layer and the medium access control (MAC) sub-layer, as well as the physical signaling (PHS) sub-layer.
- Part 2 defines the high-speed physical medium attachment (HS-PMA).
- Part 3 defines the low-speed fault-tolerant PMA.

Other parts of CAN also include:

- Part 4 defines the time-triggered communication.
- Part 5 defines the power modes of the HS-PMA.
- Part 6 defines the selective wake-up functionality of the HS-PMA.

The ISO11898-1 standard specifies the characteristics of setting up an interchange of digital information between modules implementing the CAN data link layer.[12] ISO11898-2 specifies the HS-PMA of the CAN, a serial communication protocol that supports distributed real-time control and multiplexing for use within road vehicles.[13] This includes HS-PMAs without and with low-power mode capability as well as with selective wake-up functionality.

ISO 11898-2 has been updated in parallel with ISO 11898-1 to combine the functions described in ISO 11898-2, ISO 11898-5 and ISO 11898-6. The future edition of ISO 11898-2 will cancel and replace the current ISO 11898-2:2003, ISO 11898-5:2007 and ISO 11898-6:2013.

8.7 NMEA-0183

The NMEA 0183© Interface Standard, copyrighted by the NMEA, is used worldwide across many industry segments.[14] The data bus protocol standard for serial data communication defines electrical signal requirements, data transmission protocol and time, and specific sentence formats for a 4800-baud serial data bus. Each bus may have only one talker but many listeners and is intended to support one-way serial data transmission. Data is transmitted in a series of sentences, with each sentence containing a header, a data field and a checksum. The header identifies the type of sentence, and the data field contains the actual data. The checksum is used to verify the integrity of the data. This data is in printable ASCII form and may include information such as time, position in latitude and longitude, speed, water depth, heading and distance traveled.

The IEC 61162-1 standard is a technical specification that defines the physical and electrical characteristics of the NMEA-0183 data bus. It is still widely used on ships, and it is supported by a wide range of devices including GNSS receivers, depth sounders, autopilots, chart plotters, autopilots and engine control systems.

8.8 NMEA-2000

The NMEA 2000® standard, a registered Trademark of the NMEA, contains the requirements of a serial data communications network to interconnect marine electronic equipment on vessels.[15] According to NMEA, the standard describes a low-cost, moderate capacity bi-directional, multi-transmitter/multi-receiver instrument network to interconnect marine electronic devices. Based on the CAN, it is multi-master and self-configuring, and there is no central network controller. Equipment designed to this standard can share data, including commands and status, with other compatible equipment over a single channel.

NMEA-0183 and NMEA-2000 are two different data buses that both are used to communicate between marine electronics devices. NMEA-0183 is a legacy standard that has been around for many years, while NMEA-2000 is a relatively more recent standard offering many advantages over NMEA-0183. These include higher data rates, a more efficient binary data format, a more versatile bus rather than a star topology, greater fault tolerance and greater security.

The IEC 61162-3 standard is a technical specification that defines the physical and electrical characteristics of the NMEA-2000 data bus. The standard also defines the data format and protocol that are used to transmit data over the bus.

8.9 RS-485

RS-485 is an industrial specification that defines the electrical interface and physical layer for point-to-point communication of electrical devices over long cabling distances in electrically noisy environments.[16] It is used onboard ships as a means to connect many different types of sensors for cargo and ship infrastructure monitoring. A serial communication standard that defines the electrical characteristics of a balanced differential voltage interface, it can support up to 32 devices connected to the same ISO 8482 bus. The RS-485 data bus protocol is also known as Telecommunications Industry Association (TIA)-485, Electronic Industries Alliance (EIA)-485 and International Standards Organization (ISO) 8482. Specifications include electrical characteristics of RS-485, such as voltage levels of ±5 V, ±12 V, or ±24 V; current levels and the impedance of the bus.[17] The standard specifies timing requirements, including bit rate, start and stop bits, and parity bit. It also describes the physical medium characteristics for twisted pair multipoint interconnections in either two-wire or four-wire network topology, supports binary and bi-directional signal transfer, and defines the electrical and mechanical design of the endpoint system branch cables and the common trunk cable, which may be up to 1,200 m in length. Also specified are the component measurements of the integrated type generators and receivers within the endpoint system. The data signaling rate can go up to 12.5 Mbps. RS-485 can support topologies that include point-to-point, multi-drop and multipoint configurations.

8.10 Universal Serial Bus (USB) Ports and Devices

USB ports and the devices that plug into these ports on ships connect many peripherals to a ship's computer system, including keyboards, mice, printers and storage devices. They are also used to update the software on a ship's computer system and for data transfer between a ship's computer system and other devices such as laptops and external hard drives for backing up data and sharing files. Many manufacturers of marine electronics are also incorporating USB ports into their products.

USB devices are typically "plug-and-play", where they can be connected and disconnected without requiring a system restart. The USB interface provides power to connected devices, allowing many peripherals to draw power directly from the USB connection without the need for additional power sources. Note also that the use of USB thumb drives by crewmembers and outside service and maintenance personnel poses significant exposure to cyber criminals who can use flash drives to attack the ship's computer systems and networks. The USB standard is developed and maintained by the USB Implementers Forum (USB-IF), a nonprofit organization that promotes and supports USB technology.

8.11 Bluetooth

Bluetooth is a wireless technology that allows devices to communicate with each other over short distances.[18] It performs many of the same functions as a USB port and also provides the capability to connect sensors on ships. It is also susceptible to the same risks as USB in providing access to attack ship's computer systems and networks through hacking and cyberattacks. Operating in the unlicensed 2.4 GHz radio frequency band for industrial, scientific and medical equipment, its low-power transmissions do not need to be licensed. Bluetooth is managed by the Bluetooth Special Interest Group (SIG), a consortium of various companies.

8.12 Wireless Fidelity (Wi-Fi)

Wireless Fidelity, or Wi-Fi, represents a set of standards developed by the Institute of Electrical and Electronics Engineers (IEEE), collectively known as IEEE 802.11.[19] These standards include IEEE 802.11a, IEEE 802.11b, IEEE 802.11g, IEEE 802.11n, IEEE 802.11ac and IEEE 802.11ax, which represent operations on the 2.4 GHz and 5 GHz frequency bands with speeds ranging between 2 Mbps and 1.3 Gbps. The 2.4 GHz transmits data at a slower speed than 5 GHz but does have a longer range than 5 GHz. The 5 GHz transmits data at a faster rate, but it has a shorter range because it has a higher frequency.

References

1 MSC 67/22/Add.1, ANNEX 17. Resolution MSC.64(67), 4 December 1996. Add 1. ANNEX 1. Recommendation on Performance Standards for Integrated Bridge Systems (IBS).

2 SOLAS chapter V adopted in December 2000.

3 Joint Committee Guides Metrology: Evaluation of Measurement Data-Guide to the Expression of Uncertainty in Measurement (GUM 2008), 2008.

4 H.Y. Teh, A.W. Kempa-Liehr, and K.IK. Wang. (2020) Sensor data quality: A systematic review. Journal of Big Data 7, 11. https://doi.org/10.1186/s40537-020-0285-1

5 Data Rate. PCMag Encyclopedia. *PC Magazine*. www.pcmag.com/encyclopedia/term/data-rate

6 SAE J1939 – Recommended Practice for a Serial Control and Communications Vehicle Networks. J1939 Digital Annex J1939DA_202305. Society of Automotive Engineers (SAE). 10 May 2023. www.sae.org/standards/content/j1939da_202305/

7 USB4® Specification V2.0. USB Implementers Forum, Inc. (including trademark). 30 June 2023. www.usb.org/document-library/usb4r-specification-v20

8 Ethernet Standards and Protocols Explained. Computer Networking Notes, 2023. www.computernetworkingnotes.com/networking-tutorials/ethernet-standards-and-protocols-explained.html

9 Keith Shaw. The OSI Model Explained and How to Easily Remember Its 7 Layers. 14 March 2022, Network World. www.networkworld.com/article/3239677/the-osi-model-explained-and-how-to-easily-remember-its-7-layers.html

10 Mary E. Shacklett. What Is TCP/IP? TechTarget, Networking. July 2021. www.techtarget.com/ searchnetworking/definition/TCP-IP

11 ISO/IEC 7498-1:1994. Information Technology – Open Systems Interconnection – Basic Reference Model: The Basic Model. https://webstore.ansi.org/standards/iso/isoiec74981994

12 ISO 11898-1:2015(en). Road Vehicles, Controller Area Network (CAN) – Part 1: Data Link Layer and Physical Signalling. International Standards Organization (ISO), 2015. www.iso.org/obp/ui/#iso:std:iso:11898:-1:ed-2:v1:en

13 ISO 11898-2:2016(en). Road Vehicles, Controller Area Network (CAN) – Part 2: High-Speed Medium Access Unit. International Standards Organization (ISO), 2016. www.iso.org/obp/ui/#iso:std:iso:11898:-2:ed-2:v1:en

14 NMEA 0183-Version 4.11. Serial Data Networking. National Marine Electronics Association (NMEA). November 2018. www.nmea.org/nmea-0183.html

15 NMEA 2000® Version 3.000. The Modern Marine Interface Standard. CAN Based Serial Data Communications. Vessel Systems Integration. April 2022. www.nmea.org/nmea-2000.html

16 Jason Kelly. RS-485 Serial Interface Explained. CUI Insights. CUI Devices. www.cuidevices.com/blog/rs-485-serial-interface-explained

17 ISO/IEC 8482:1993. Telecommunications and Information Exchange between Systems, Twisted Pair Multipoint Interconnections. International Standards Organization (ISO). www.iso.org/standard/20954.html

18 Bluetooth Specifications and Documents. The Building Blocks of All Bluetooth Devices. Bluetooth Special Interest Group (SIG). www.bluetooth.com/specifications/specs/

19 IEEE 802.11-2020. IEEE Standard for Information Technology Telecommunications and Information Exchange between Systems. Local and Metropolitan Area Networks: Specific Requirements. Part 11: Wireless LAN Medium Access Control (MAC) and Physical Layer (PHY) Specifications. Institute of Electrical and Electronics Engineers, 2020. https://standards.ieee.org/ieee/802.11/7028/

Chapter 9

Cybersecurity and Ship Sensors

During a conference in 2020, Robert Rizika of the NavalDome cyber defense company reported that cyberattacks on maritime industry operational technology (OT) systems increased by 900% from 2017 to 2020, with many more attacks going unreported and the trend continuing to increase.[1] Reported cyberattacks among maritime industry stakeholders have increased from 43% in 2018 to 74% in 2022, according to Ford Wogan of the Jones Walker's Maritime Practice Group.[2] The International Association of Ports and Harbors (IAPH) has been quoted as stating the maritime industry globally suffered a fourfold increase in cyberattacks between February 2020 and May 2020 alone.[3] This is a significant issue that is ever-increasing in importance and impact on maritime operations.

Many intrinsic cyberattack vulnerabilities must be overcome in data, network hardware, software and sensor-intensive critical ship infrastructure utilizing communication networks that service transportation systems. The specific focus of this chapter is on ships and by extension, the ports they call upon where cyberattacks were identified by the U.S. Secretary of Homeland Security as "the most significant threat to port infrastructure".[4] Disruptions related to cyberattacks can affect tens of thousands of ships and thousands of seaports, many of which are vital to energy infrastructure. Cyberattacks can, and often do, propagate throughout the worldwide logistical supply chain. This problem will continue to increase in importance as port digitalization expands through greater integration of smart devices and Internet of Things (IoT) technology that often includes wireless communications.

The hundreds to thousands or more sensors onboard a typical ship are of diverse vintages and technologies and are designed and manufactured by many different companies. Sensors are maintained by even more companies and operated by a third group of companies; all following different business models and security practices. Smart and IoT sensors further complicate the problem by promoting widespread connectivity that often takes shortcuts across multiple security layers to perform their jobs. Yet the unique characteristics of numerous and diverse sensors and the wide range of functions and applications they serve are often ignored and relegated to a level of lesser

DOI: 10.1201/9781003377900-9

importance than cyber systems in general. Indeed, sensors are designed to share data, but sensor communications are often unencrypted and fairly easy to intercept. This is a mistake.

Maintaining safe and secure critical infrastructure across the sensor environment is essential to ensure efficient, uninterrupted and uncompromised vessel operations. This issue encompasses onboard sensors and extends throughout the entire sensor ecosystem, including off-vessel users, company offices and remote control centers. This also extends to government agencies with which ships interact to fulfill regulatory requirements and the port facilities that comprise the land-ship interface where routine and essential ship operations take place.

The challenges of ensuring sensors remain operational, viable and free from interference and disruption embrace many of the same principles devised for cybersecurity in general. The complexity of cyber-physical-data critical sensor infrastructure can also hide vulnerabilities to small and seemingly unimportant changes and variations within their makeup that can make them susceptible to catastrophic and destabilizing behavior such as cascading failures and open back doors to infiltration. The ability to detect such sensitivities and secure against attacks to trigger them is a key factor when implementing effective sensor cyber security.

Cyberattacks on large information technology infrastructure often make headlines when the identities and personal data of millions of people and organizations are stolen. Likewise, the immediate needs of communities, businesses and households are adversely affected when critical operational technology, such as electricity generation and energy distribution infrastructure, is compromised. However, the effects of sensor compromise may often be subtle, and long-term and manifest themselves indirectly as a result of breakdowns and similar events that are hard to trace to their source. A prime example occurred in 2010 when thousands of centrifuges at the Natanz fuel enrichment plant in Iran were destroyed. This was accomplished through the infiltration of computer systems by the Stuxnet computer virus that changed sensor signals to make it appear operations were nominal yet the centrifuges were actually spinning far above their normal speed and tearing themselves apart.[5]

Similar analogies may be made to shipboard sensor systems. Some examples of the possible consequences of cyberattacks include the ability to:

- Incapacitate onboard computers to make sensors inoperable or blind.
- Spoof GNSS signals on cue to ground a vessel in the middle of a critical waterway.
- Corrupt electronic navigation charts with false data and soundings to obscure hazards to navigation.
- Hijack ship and vehicle controls.

- Implant descriptions of unauthorized people within smart cameras using facial recognition to provide crewmember and passenger access to restricted areas.
- Disable engine pollution monitors and controllers, resulting in spills and discharges of hazardous chemicals and pollutants with great environmental impact.
- Degrade the resolution and effectiveness of cameras and other sensors.
- Render object and event recognition capabilities of sensors ineffective or useless.
- Weaponize civilian ships for nefarious purposes.

Areas of commonality between information technology (IT) and OT systems and cybersecurity, in general, are described to provide context and to draw attention to topics unique to ship sensor functions and installations. The nature of cyberattacks is also covered providing examples of how the attacks are committed and the potential consequences of these attacks. Also discussed is the muddy grey cyber line that distinguishes where the ship ends and the port begins, especially in current times when constant contact is routinely maintained between ships and shoreside organizations through satellite and terrestrial 5G communication links. Details are provided regarding critical ship and associated port infrastructure that require protection and the means through which this may be accomplished. The chapter concludes by citing resources describing precautions that may be taken to help minimize the potential for cyberattacks on sensor systems and to limit their severity when they occur.

Please note that while this discussion specifically addresses sensor systems onboard ships, many of these same principles apply directly to a wide variety of sensor applications and related networks that span the maritime industry throughout the ports themselves and to transportation in general. These concepts can also be readily adapted to meet sensor cybersecurity needs of telecommunications, energy production and distribution, pharmaceutical, medical and many other industries.

9.1 Aspects of Cybersecurity Unique to Sensor Systems

Cyberattacks against shipborne sensor systems often take a fundamentally different form than many types of attacks routinely committed against computer networks in general in terms of their form and function, and the consequences of their occurrence. Sensor networks, by their very nature, are likely to have relatively few active human users and real-time interactions with other Information Technology (IT) and Operational Technology (OT) systems, and they should operate through one or more firewalls that can detect the legitimacy of access requests. They should also be well-segmented and isolated by virtue of access being physically or electronically limited to only a relatively few people for maintenance and system upgrades. Vulnerabilities encountered through outbound communications from a sensor network are also likely to be greatly minimized as the destinations for sensor data and analytics products should be well-defined and restricted. These characteristics significantly hamper several of the more prominent types of direct cybersecurity attacks, such as phishing for data through malicious emails, attacks resulting from visiting legitimate websites with weak security that are infected, and clone websites pretending to be legitimate that are also infected with executable scripts. Brute force attacks attempting to guess login credentials are also not likely to succeed since multiple attempts to gain access can easily be detected and cause the user to be locked out. Also unlikely are denial-of-service (DoS) attacks originating from external IT sources that attempt to flood and overwhelm a sensor network server with traffic until it can no longer respond or crashes, thereby preventing access to authorized users. However, vulnerabilities to DoS attacks still exist via the sensors, the people who maintain them, and the crews and passengers that are present on the ship.

Much more likely to occur are insider attacks resulting from the introduction of malware through software and firmware updates accomplished using USB devices as well as "trusted" users whose end-point systems are themselves compromised. Such circumstances allow viruses and other malicious code to be passed along to computer server(s) and/or sensor controller(s) that interact with the server network and directly with the sensors. The malware may act immediately upon insertion or may lie dormant for long periods of time, awaiting the occurrence of predefined circumstances before acting. These circumstances may include a combination of unique operating conditions exhibited by an engine or other ship's machinery, a specific date and time, a vessel's position defined by latitude and longitude, and recognition of known unique visual or digital signature(s) introduced externally to the ship through an onboard camera.

Opportunities for breaching security by monitoring data communicated between sensors and their controllers and between the controllers and the server(s) are also made available by middleware that functions as a hidden layer between an operating system and sensor software applications. Middleware increases the possibility to insert ransomware that targets known vulnerabilities, or malicious software that models the sensor environment to devise methods on its own to detect and attack vulnerabilities within the network itself and/or the individual sensors operating within the network.

Another problem in maintaining cybersecurity is the need to ensure conformance and interoperability of methods and processes to provide users with resilient and high-performance network capabilities that include quality of service, prioritization and the avoidance of service pre-emption from both accidental and nefarious sources. Cybersecurity technologies are relatively new and many products offered by the various vendors and different service providers often exhibit poor communication and interoperability at both the endpoint applications and at the servers that manage these applications. Such conditions provide multiple vulnerabilities and opportunities to infiltrate, disrupt and otherwise compromise sensor operations.

9.2 Cyberattacks and Sensor Systems

Several examples of cyberattacks on sensors used onboard ships are described in this paragraph. It must be noted that other industries more advanced in using smart sensors and automation technology are also experiencing this same problem. Examples are cited from which lessons can be learned and ships can be protected in future practice. An attempt is made to provide coverage across a range of sensors and sensor types most commonly found on ships and maritime vehicles. These examples are provided to initiate discussion through which further action may be taken to help ensure incidents such as these do not occur or are minimized in their effect.

9.2.1 Engineering Sensors

Various engineering functions and measurable parameters discussed in Chapter 2 are rapidly evolving with the development in the commercial marketplace of vessel control systems for local and remote monitoring and control of ship infrastructure such as engines and propulsion, generators, batteries, steering, environmental monitoring, alarms and many other systems throughout the vessel. These systems form the core OT to operate the vessel itself, and any attempts to interrupt their normal operations may place the crews, passengers, cargo and the ship itself in jeopardy. The following incidents and events help to illustrate some essential points on this topic:

- A plan was identified to interfere with normal seawater ballast operations necessary to maintain ship proper trim, stability and structural integrity.[6] This included a scheme to take control of the ballast pumps remotely by sending commands via a satellite link to cause problems that could result in the sinking of the ship or other significant and irreparable damage to the vessel.
- Vulnerabilities in a ship gas turbine power control system were revealed by several methods through which it was possible to gain access to and tamper with various control parameters.[7] The scenario identified an onboard distributed control system connected to a remote center via a satellite network. This single communications link was connected to a complex network structure with a large data flow that was extremely vulnerable to several cyberattack methods. These included viruses and DoS attacks that could result in abnormal temperature and pressure changes in the gas turbine and may lead to damage to system components and the

complete shutdown of the turbine. This study ultimately resulted in the development of a method for risk assessment that can detect the vulnerabilities, identify potential attack mechanisms and take the appropriate defensive measures.

- A comprehensive analysis of vulnerabilities to systems responsible for ship power generation and propulsion equipment identified numerous threats of cyberattacks, many of which can result in a high impact on ship operations that were also highly likely to occur.[8] These include falsifying data regarding engine lubrication, tampering with engine commands, interruption of engine automation system operations, data integrity violations that can cause malfunctions in critical operations such as rudder control, alteration of fuel consumption data, falsification of equipment maintenance data and compromise of handling systems resulting in damage or loss of cargo and interruption of emergency handling processes.
- Several types of vessels, including container ships, tankers, and cruise ships, were examined to illustrate the ease with which taking over control of various ship's systems could be accomplished. The participants in this experiment were able to disable entre systems as well as compromise engines, pumps, machinery, steering gear controls, overwrite signals coming from ballast sensors, pumps and fuel level sensors.[9]

9.2.2 Navigation Sensors

The complete range of navigation sensors described in Chapter 3 are also valuable targets for cyberattacks due to their essential nature in ensuring the safety of navigation. The literature is filled with examples of how potential and actual attacks on sensors can at best create confusion that may require immediate corrective action on the part of watchstanders; and at worst cost lives, loose cargo and sink ships with dreadful environmental and supply chain consequences. The following are just a few examples describing cyberattack attempts and vulnerabilities with navigation sensors.

- A comprehensive cyber risk assessment was performed to determine the vulnerability of the ship's Electronic Chart Display and Information System (ECDIS).[10] An outdated Windows operating system was discovered on the computer where ECDIS was installed, providing little protection against cyberattacks.
- A report of efforts to compromise ship systems described their ability to attack a ship's navigation system and radar and take over the steering of the ship.[11] System penetration was achieved by sending an email to the captain's computer that enabled access to ECDIS. An attack was simulated at night through a narrow canal altering the ship's course while continuing to provide normal instrument display indications to the Officer of the Watch, effectively running the ship aground.

- Additional information was obtained regarding the above experiment that targeted the ship's radar system using the Local Area Network (LAN) that is also connected to ECDIS, the bridge alert system and the voyage data recorder (VDR).[12] The results included being able to eliminate radar targets from the screen while the system appeared to be working perfectly.
- Numerous attacks on Global Navigation Satellite System (GNSS) services have been reported worldwide, particularly in and around Russia, Korea and in the Mediterranean, Red and Baltic Seas. These efforts have resulted in DoS and spoofing incidents that render satellite positioning indications useless or worse, deceptive so that a ship could be lead astray for nefarious purposes.
- A spoofing attack was made on a 213-foot yacht using the U.S. Global Positioning System (GPS) version of GNSS in the Mediterranean.[13] The experiment took place approximately 30 miles off the coast of Italy while en route from Monaco to Rhodes, Greece, as the yacht sailed in international waters. False GPS signals that were not distinguishable from authentic signals covertly overpowered the ship's navigation system until control of the ship's navigation system was achieved. Subtle maneuvers were then made to position the yacht a few degrees off its original course to follow a parallel track hundreds of meters from its intended one. Throughout this time, the chart display showed a straight-line path being followed while the ship's wake formed a pronounced curve, showing that the ship had turned.
- Communications with a U.S. RQ-170 drone aircraft were jammed by Iran, where they were also able to reprogram their GPS coordinates to land in Iran and be captured rather than land at their intended location in Afghanistan.[14] This same exact scenario could also be used to hijack a ship and divert it into the territorial waters of a hostile country or the hands of pirates.
- Automatic Identification Systems (AIS) have traditionally been easy to hack and are a primary sensor targeted when attempting to disguise ship locations and information. AIS broadcast data includes vessel type, name, position, course and speed, along with other information useful to nearby vessels, vessel traffic services (VTS), and port authorities. It is communicated using unencrypted Very High Frequency (VHF) radio signals lacking internal authentication and integrity checking of any data.[15] Pirates often use AIS tracking to target certain vessels and carry out malicious activities, including shutting the system off while underway to hide their presence. There are many incidents where the physical locations of ships did not correspond with their positions as indicated on AIS, and were off sometimes by hundreds or even thousands of miles. The problem is especially prevalent with naval vessels, where between August 2020 and July 2021, almost 100 naval vessels with suspected false AIS tracks were detected.[16] This is also a major problem with commercial vessels.

- VDRs are targets for attack as they record data from multiple sensors, including speed, direction, position and radar images in addition to audio recordings and other data that are indispensable to help identify the cause of maritime incidents. One incident was reported in India where VDR data from a cargo ship that hit a fishing vessel was erased after an infected USB stick was inserted to retrieve the data.[17] An analysis of another VDR cited in the same source also detected weak encryption, an insecure mechanism for user authentication and various service vulnerabilities.[18] This could lead to modifying data to, for example, delete audio conversations from the bridge, delete radar images, or alter speed or position readings, and even spy on a vessel's crew as VDRs are directly connected to microphones located on the bridge.[19]
- Video cameras are appearing on many ships as an essential means to monitor engine rooms, cargo holds, closed spaces and other locations that are generally unoccupied and sensitive areas where security is an issue. Many of these cameras are connected to networks and are IoT devices containing vulnerabilities that could let attackers watch live camera feeds, create botnets, or use hacked devices as a stepping stone to further attacks.[20] Another problem is that many cameras used on tankers, grain carriers and other potentially hazardous surroundings do not meet safety requirements for use in explosive environments.[21] Such characteristics can be exploited to cause abnormal operations that may overheat the camera and cause an explosion.
- Satellite communications have traditionally been a high-value target for attack. Weak encryption and old IT equipment pose key vulnerabilities.[22] Also, if an individual IoT device is not encrypted or every stage of data communication is not properly protected, an entire network of connected devices can be manipulated. Several incidents identified in this paragraph originated through the interception and manipulation of satellite communications. Key components to satellite security are the ground networks that move data and links that transmit and receive satellite data through which back door assaults can take place.[23] Satellites are not useful if their access and performance are not assured. This has been a significant tactical target for both sides fighting the war in Ukraine.
- Maritime Very Small Aperture Terminals (VSATs) used for two-way real-time data communication via satellites present cyber threats to vessels due to the value of the data they transmit and their role as attack vectors by providing access to vessel IT and OT infrastructure.[24] Cyberattacks on vessels can be executed by discovering and exploiting vulnerabilities in a sequential manner by first penetrating telecom equipment and taking control of the IT segment, finding problems in segmentation and accessing the OT layer and controlling OT equipment and arbitrarily intercepting and modifying Transmission Control Protocol (TCP) sessions enabling man-in-the-middle and denial of service attacks against vessels at sea.

- Sensor data busses such as those conforming to NMEA 0183/2000, CAN and other standards are susceptible to both physical attacks by insiders as well as software attacks that can disable them and cause the sensors and displays to freeze and/or go black.[25] A problem with the bus itself may cause every instrument dependent on the bus to malfunction. Attacks may be accomplished by physically connecting to available ports (e.g., USB, PS2, serial, etc.) to insert malware, using stolen credentials, scanning the network topology, physically tampering with a device and various other means.[26] Spoofing wireless communications using Wi-Fi, sniffing network traffic between sensors, deleting files from a compromised sensor or controller and sending improper commands are several methods used to achieve digital attacks on instrument buses.
- Environmental sensors are being integrated into many current unmanned and autonomous maritime surface vehicles and future manned vessels that are presently in use and being tested on autonomous trucks, busses and automobiles. These include Light Imaging Detection and Ranging (LIDAR), millimeter radar (mmRadar), acoustic (e.g., ultrasound) and other types of sensors. Numerous attempts to attack such systems have been detected and countermeasures developed that will prove useful in future ship cyber defenses.[27]
- Weather sensors generally do not garner a great deal of press coverage as targets for cyberattacks, but their significance onboard ships cannot be overstated. This is especially true when automated ship control systems use wind velocity, direction and other factors to help maintain a stable keel while underway. Weather sensors now incorporate many new types of technologies without moving parts to perform their function when compared to past-generation devices and are often IoT-capable. Such sensors are especially susceptible to ultrasound and laser interference, and they can be exploited as a target to create an entry point to infiltrate onboard networks.

9.2.3 *Crew, Passenger and Cargo Sensors*

The list of shipboard sensors can be extended to those used to monitor crew and passenger access and locations, and cargo as described in Chapters 4 and 5. The principles and tactics associated with cyberattacks on these sensors are similar to those used against engineering and navigation sensors.

- Facial recognition cameras, fingerprint readers, iris scanners, and other biometric security methods are increasingly being used to limit crew member and passenger access to secure places onboard ships. These devices have been shown to be vulnerable to several modes of attack.[28] These include attacks via USB and other active ports to update device firmware and insert malicious software, and accessing a network that leads to the IP camera device to monitor network traffic between the device and the server. Also found possible was the ability to impersonate a connected access control device and send the server data it would recognize and validate.

- Discrete and IoT temperature, smoke, explosive vapor, and other sensors placed within inhabited, cargo, and sealed spaces are often interconnected via Bluetooth, LAN, and other networks that suffer many of the same vulnerabilities as all networks in general and can be used as gateways to infiltration. However, additional precautions need to be taken to ensure physical obstructions are not placed in close proximity to sensors so as to inhibit their normal operation by isolating them from the properties being sensed and interrupting network connections by reducing signal strength. The placement of sensors beyond their useful range has also resulted in service interruptions and disconnections from the network causing unreliable temperature measurements.[29]
- Cargo container monitoring criteria have changed significantly since 2019 with the adoption of updated sea carrier Minimum Security Criteria that require extensive use of security cameras, vessel monitoring and other requirements and best practices.[30] Many of these strategies are implemented through the use of smart sensors and IoT devices subject to the same vulnerabilities previously cited. This can help facilitate attacks that can eliminate individual containers from being tracked by corrupting databases, disguising their movements by corrupting GNSS signals and placing them in false locations, and changing their contents on manifests.
- The use of infrared cameras and perimeter Sonar may help to detect and track adversaries who attempt to overcome security precautions and evade ship defenses. The heat signature of a human in the water can be easily detected using infrared cameras, although the intensity of the signature may be reduced somewhat as a result of the swimmer wearing an exposure suite. Sonar can detect a swimmer on the surface by day or night, someone swimming just below the water's surface using snorkel gear, a diver approaching underwater from depth using open or closed-circuit breathing apparatus, as well as diver delivery vehicles. Several different manufacturers offer sonar products of this nature that generally have a range of up to 1,000 meters at depths of 50 meters.

9.3 Where the World Ends and the Ship Begins

As ships become more automated and the number of onboard sensors increases, their vulnerability to cyberattacks dramatically escalates while in port and exists as part of the port through the establishment of direct physical and electronic connections with port infrastructure. This is due primarily to continuing port automation that also increases network vulnerability as more equipment and sensors are added and with greater use of remote-controlled port machinery such as gantries, cranes and bridges. Another major contributor to the problem is access to the ship by technicians, maintainers and port employees. Nefarious interference with port operations, systems, apparatus and devices comprising port infrastructure can result in death or injury to port employees, ships' crews and passengers; destruction of property and cargo, loss or compromise of sensitive data, extended cargo delivery schedules and/or environmental hazards. This is aggravated as a result of attack strategies that can evade the present generation, less than real-time detection capabilities of disruption to landside and shipboard facilities. The problem extends to port state control as implemented through port control systems (PCS), cargo control systems (CCS), and related means to connect multiple systems across different government, public and private organizations for the secure and intelligent exchange of information.

Cybersecurity must not only focus on ship infrastructure, but also on port of origin and destination segments of the entire shipping process. Figure 9.1 illustrates these segments and the additional shoreside components and processes that can be vulnerable to cyberattacks through physical and electronic methods.

This cybersecurity environment is very complex and consists of multiple segments related to logistics, engineering, security and management elements representative of cargo transported, compliance, safety and other functions. This also includes physical port and vessel facilities in terms of cyberinfrastructure and sensor-based systems through which maintenance is performed, security is sustained along with many other operations, and upon which ships rely to safely and efficiently transport cargo between ports. By focusing specifically on only the ports and ships themselves, rather than the entire shipper-to-consignee process, including handling, haulage and warehousing; the problem is reduced to a more manageable size and complexity.

The physical interfaces between the port and the ship include wire, fiber-optic cabling and hoses to provide electrical power, communications and utility services in addition to gangways and ladders across which are transported physical sensors, new and replacement parts and systems, and various media to update network and sensor systems by technicians and maintainers.

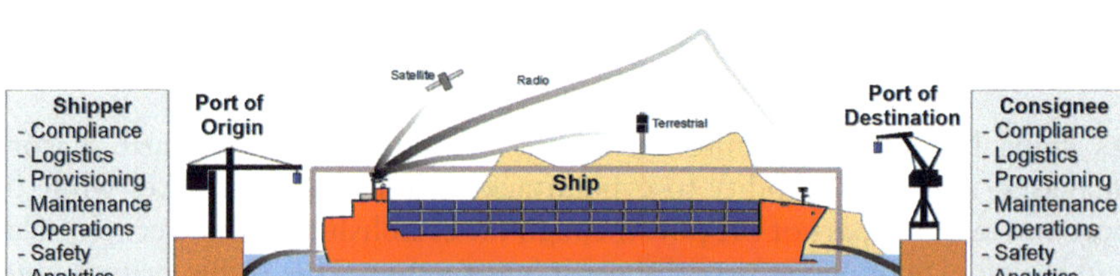

Figure 9.1 Port and Ship Infrastructure Segments of the Entire Shipping Process.

This also includes any devices and equipment brought onboard by crewmembers and passengers, and devices that may be embedded into cargo and/or their containers.

Digital interfaces while the ship is in and close to the port include Wide Area Network (WAN) and wireless communications provided by the port itself and its immediate vicinity by the local community. In addition, all shipboard capabilities in terms of satellite, radio and terrestrial cellular communications continue to be available. Significant changes occur in both the port and ship cybersecurity environment when shoreside connections are slipped and the ship leaves port, and when a ship arrives in port and shoreside connections are again established. The only remaining shipboard capabilities that should exist can be limited to satellite, radio and terrestrial cellular communications, as well as cellular devices in the personal possession of crewmembers and passengers. However, there are exceptions.

While onboard sensors are operational, they can be subject to a wide variety of external stimuli that provide unconventional and unintended, two-way access to the ship's digital infrastructure. These sensors include cameras, weather instruments and many other devices accessible via light, laser, acoustic, microwave, radio frequency (RF) sources and other means of stimulus that may contain data, information and intelligence embedded within their transmissions. All such means of access must be considered as part of the infrastructure.

9.4 The Attack Surface, or the Protect Surface?

Much of the literature discussing cybersecurity has traditionally discussed the *attack surface* against which cyberattacks are made, and the need to reduce this to as small a footprint as possible. One definition describes it as the total number of all possible entry points for unauthorized access into any system, including vulnerabilities and endpoints that can be exploited.[31] This also represents the entire area of an organization or system that is susceptible to hacking. The smaller the attack surface, the easier it is to protect. However, most maritime infrastructure is already massive and as new technologies are introduced the attack surface continues to expand. Also, with the increasing use of IoT devices and sensors the attack surface has expanded exponentially.

The U.S. Department of Defense (DoD) has recently released its *Zero Trust Strategy and Roadmap,* which anticipates current and future cyber threats and attacks that go beyond the traditional perimeter defense approach.[32] Rather than looking at the attack surface from a high level, in Zero Trust, the exact nature of what is needed to be protected is defined as the *protect surface*, which is the smallest possible reduction of the attack surface.[33] It is defined based on the data to be protected, application usage of sensitive data, asset vulnerability and services that can be exploited to disrupt operations. The protect surface is orders of magnitude smaller than the overall attack surface and it is always knowable. Firewalls and other controls are moved as close as possible to the protect surface rather than the perimeter at the attack surface, where it is decidedly further away from what needs to be protected. This is much like identifying the protect surface as a person's lungs to place a "firewall" to battle lung cancer rather than defining the attack surface as the skin at the body's perimeter to keep the cancer from entering the body itself (probably a bad analogy, but I am not a medical doctor so I will use this as my excuse!). In this way, it is possible to determine what traffic moves in and out by a very small number of users or resources that actually need access to sensitive data or assets. As with the attack surface, organizations must constantly monitor their protect surface to identify and block potential threats as quickly as possible. However, in theory, the smaller footprint makes this process more manageable.

Actual methods to model the protect surface are still being developed. With implementation of distinct DoD Zero Trust capabilities and activities anticipated by 2027, there remains little time to determine how this will actually be accomplished.

9.5 Ship and Port Cyber Infrastructure

It is necessary to create a model to identify, assess and specify attributes and characteristics comprising port and ship infrastructure to be able to assess what actually must be protected. The focus should be on the design and implementation of the overall architecture through which networks characterize cargos, sense and perceive the physical port and maritime environments as well as internal capabilities to fulfill network requirements. This includes how network data and the physical and environmental characteristics they represent are understood, and the fusion of these data to achieve a goal of comprehensive, continuous, near-real-time (in fractions of a second) assessment of potential threats through reduced-order and requirements-based variability modeling. Considering the contrasts between the concepts of attack surface and protect surface modeling described in the previous paragraph, it is easy to see the differences in the magnitude of effort required to perform these tasks.

The ship and port physical and digital infrastructure that interfaces with the ship must be considered part of one entity while the ship is in port. However, the scope of infrastructure decreases dramatically when the ship gets underway and the number of entry points to the ship are greatly reduced.

9.5.1 *Ship Infrastructure*

Computer networks onboard ships are generally comprised of the same types of equipment and software found in ports. They are tailored to the specific requirements needed to perform their missions in addition to moving their passengers and cargo from port to port wholly within the confines of the hull, deck and superstructure. However, unique to the ship architecture are computer systems to support sensor activity that should be segmented to isolate these functions from general IT operations.

Activities are undertaken through which data, information, algorithms and software updates pass to and from company IT offices, ports, government offices, regulatory agencies and the ship. These activities include freight and customs, maintenance data and diagnostics, navigation information, ship sensor data, software and software patches, middleware, and firmware.

9.5.1.1 Ship Physical Infrastructure

The physical ship infrastructure is comprised of the computer networks and all the devices and interfaces through which access to ship network systems may be achieved. There is great commonality between ship and port IT systems functions including servers, computers, laptops, mobile devices and tablets, modems, routers, switches, controllers, peripherals including disk drives, memory storage devices and monitors; printers, speakers, sensors and cameras. Segmentation of shipboard networks can often be found by function based upon cargo loading/offloading and management, engineering, propulsion and maintenance; bridge communications and navigation, administrative and crew welfare, access control, passenger services and management, and private and public networks. Included are physical interfaces that are used by maintenance and IT teams to download maintenance data, voyage data, sensor data and other information to laptops that are also used to update middleware and firmware.

Significant differences exist between port and ship OT encompassing sensors and equipment used in the engineering department. Engineering systems cover many functions with unique modalities and characteristics associated with engine, propulsion, steering, electrical, heating, ventilation and air conditioning (HVAC); distillers, condensers, evaporators and associated pumps, valves and piping. Control and health monitoring of engineering systems distributed throughout the vessel is accomplished through the use of networks of actuators and sensors to monitor and generate alarms resulting from out-of-tolerance and abnormal conditions. Sensor networks are integral to the monitoring of engines and drive trains, bearings and other components that consider temperature, wear and vibration, water infiltration into oil, as well as monitoring exhaust gas and cylinder liners. New sensor technologies also exist for level sensing, water ingress, pressure monitoring, deck winch torque and load monitoring, battery management systems and many other engineering sensing applications.

Deck department systems forming part of the physical infrastructure include the devices, monitors and peripherals associated with ECDIS, depth and navigation sonars, GPS/GNSS, AIS and satellite AIS, digital radar/

Automatic Radar Plotting Aid (ARPA), broadband satellite communications, high frequency (HF) and VHF communications, digital select calling (DSC), weather sensors, navigation and security visible and infrared cameras, navigational text messaging (NAVTEX) and autopilots.

The physical infrastructure has greatly expanded as passenger ships integrate into voyager cabins fire and smoke detectors supplemented with floor, wall and shower sensors to detect occupancy and movement. Facial recognition devices are used to open cabin doors, and wearable devices containing passenger- and crew-unique identity information are read by thousands of sensors both on board the ship and in port to enable hospitality and victualling services as well as speed the boarding and debarkation processes.

Also forming part of the physical infrastructure are shipping containers outfitted with sensors to help track movements when repositioned on ships, within ports and during transit from their place of origin to the destination. Sensors also help detect and predict conditions at sea conducive to container loss overboard and monitor the environment and detect excessive heat. Heat and smoke detectors are widely employed on many Roll-On/Roll-Off (Ro-Ro) vessels and ferries to detect and isolate smoldering and combustion among vehicles being carried. Tanker ships and gas carriers utilize special sensors to monitor levels, pressures, gas buildup and to detect foreign materials within their cargo. Bulk carriers feature sensors to detect and possibly help prevent cargo liquefaction by monitoring temperature and humidity levels both in their holds as well as within the cargo contents.

Underlying many onboard networks are digital communication buses for data transfer between sensors and onboard computers. Internal ship communication is accomplished using one or more data buses that include local area network (LAN), RS232/RS485, NMEA-0183/IEC 11623, NMEA-2000/IEC 61162-3, controller area network (CAN) and Bluetooth; with communications beyond the confines of the ship accomplished through 5G terrestrial networks, wide area network (WAN) and satellites. Limited data transfer is also performed using VHF radio for AIS data between ships and vessel traffic control centers. It should also be noted that digital communications protocols are also now being used on a limited basis for data, telemetry and image exchange using HF radio primarily as a backup for long-distance satellite communications.

9.5.1.2 Ship Digital Infrastructure

Digital ship infrastructure is comprised of all of the software, middleware, firmware, and related mechanisms reachable by the outside world through which access to ship network systems may be achieved. The digital elements of ship IT infrastructure can be summarized as mirroring those of port infrastructure counterparts as tailored to the specific requirements of the ship and its mission(s). This is discussed in the paragraph for Port Digital Infrastructure.

However, ship OT digital infrastructure differs greatly from those of ports in terms of functionality and data content unique to support the needs of physical components serving engineering, deck, passenger, cargo and other relevant operations. This varies widely depending on the type of ship involved and its function.

9.5.2 Port Infrastructure

Ports of origin as well as destination ports serve as primary focal points where the vast majority of physical and electronic interactions with ships occur. Port activities are represented in the physical and digital infrastructure necessary to accomplish all functional aspects of ships' voyages. This includes the loading and offloading of cargos, passengers and crews and their corresponding records and documentation; transmission of bills of lading, logistics, customs and compliance data and the gathering of operational and maintenance data from preceding voyage(s) and legs of a voyage. Also included is the repair and installation of physical equipment at the port itself, shipboard deck and engineering equipment as well as computers, sensors and operational equipment, software, middleware, and firmware; provisioning and victualing supplies. Our approach examines port physical and digital infrastructure as separate and unique from ship infrastructure yet comprising a hybrid ship/port infrastructure through which the vast majority of interactions with the ship take place. This comprises a complex and varying infrastructure that also makes the ship highly vulnerable while in port.

9.5.2.1 Port Physical Infrastructure

The physical port infrastructure is comprised of all of the devices and interfaces visible to and accessible by the outside world through which access to port and ship network systems may be achieved. This includes a broad range of port equipment and devices related to IT and OT regardless of whether it is physically located at the port itself or remote from the port yet connected via electronic means. The breadth of the physical infrastructure spans multiple levels from the largest of cranes and gantries used to load and unload ships' cargos; to the buildings that house the servers, computers, personal computers (PCs), firewalls modems, routers, switches, memory devices, clients, IoT endpoints, computer and other electronic IT ports, cables and wireless connection paths that process data and information related to port operations; as well as the entry points, security devices, cameras and sensors associated with these buildings and surroundings. Also included are Internet web portals that allow shippers and consignees to track cargo, check pickup availability, review ship schedules, file bills of lading and many more functions. Physical infrastructure also extends to the cargo containers outfitted with smart sensors monitoring their contents and movement of the containers themselves, and to smart sensors placed within and monitoring the conditions of bulk cargos, including programmable logic controllers (PLCs) and arrays (PLAs) integral to remote control and smart sensor designs. Also considered are the systems used to process the cargo, passengers and crews, maintenance, data acquisition, processing and analytics across various commercial organizations and governmental agencies performing both direct and supporting roles within the port and onboard ships.

Their configurations may be characterized in terms of physical assets, the functions they perform, how they are interconnected, the means by which they are accessed and other pertinent details. This includes noting specific details by equipment types and models, capabilities, accessible ports and the means used for access and update, including human access using physical (laptop, mobile devices, USB stick) and electronic (LAN, WAN, Bluetooth, etc.) methods.

9.5.2.2 Port Digital Infrastructure

Digital port infrastructure is comprised of all of the software and related mechanisms reachable by the outside world through which access to port and ship network systems may be achieved. Digital elements include operating systems, application software, middleware, firmware, utilities, device drivers, databases, enterprise software, application suites and communications software. Additional elements include IoT endpoint software, mobile and web applications, and websites such as iDockworks, Navis N4, TOS Web Portal and VoyagerTrack;[34] remote desktop protocol (RDP) endpoints, email inboxes, virtual private networks (VPNs), cloud services as well as supply chain infrastructure and services. Represented within these elements are a wide variety of data and data types, communication protocols, implementation software and the communications capabilities needed to perform routine and extraordinary company operations in terms of services and the transportation of passengers, goods and materials on a day-to-day basis. Also included are capabilities for handling security, regulatory, safety, maintenance, logistics, passengers and crews, voyage planning, voyage and cargo tracking, victualing and other human needs provisioning under nominal and abnormal conditions and circumstances. Significant attention must be given to the use of cloud services that produce a very large digital infrastructure and are managed very differently than on-premise storage, change dynamically and can easily compromise security. Also needed to be considered is the effect on port digital infrastructure resulting from digital twin technology used to duplicate all port software functions in a virtual representation of operational systems.

9.6 Present-Day Precautions Against Cyberattacks

Many excellent resources are available to assist in establishing cyber security onboard ships, in ports, and in IT and OT infrastructure in general. A good starting point is the International Maritime Organization (IMO) itself, which specifically addresses and provides guidance for maritime cyber risk and risk management. Additional guidance includes publications and other resources available from the Baltic and International Maritime Council (BIMCO), American Bureau of Shipping (ABS), Digital Container Shipping Association (DCSA), International Association of Classification Societies (IACS), International Association of Ports and Harbors (IAPH) and the United States Coast Guard. Much of the guidance and many of the recommendations that have been developed for maritime use are guided by the U.S. National Institute of Standards and Technology (NIST) National Cybersecurity Center of Excellence in the development of their *NIST Cybersecurity Framework*. The contributions of each of these organizations are summarized in the following paragraphs.

International Maritime Organization (IMO) – The IMO has issued guidance and adopted a resolution to address maritime cyber risk, which may result in shipping-related operational, safety, or security failures as a consequence of information or systems being corrupted, lost, or compromised. *Guidelines on Maritime Cyber Risk Management* (MSC-FAL.1/Circ.3) provide high-level recommendations to help safeguard shipping from cyber threats and vulnerabilities that thwart effective cyber risk management.[35]

The IMO Maritime Safety Committee in June 2017 also adopted Resolution MSC.428(98), *Maritime Cyber Risk Management in Safety Management Systems*, to encourage national administrations to ensure cyber risks are appropriately addressed in existing safety management systems no later than the first annual verification of the company's Document of Compliance after 1 January 2021.[36]

Baltic and International Maritime Council (BIMCO) – In 2018, BIMCO, in cooperation with other maritime industry organizations, released their *Guidelines on Cyber Security onboard Ships* based on high-level principles.[37] The update to this document issued in 2020 with its 4th edition seeks to establish safety, security and commercial risk awareness that results from a lack of cyber security measures. Topics considered include the protection of shipboard IT infrastructure and connected equipment, a system for authentication and authorization of users, protection of data used in the ship environment, management of IT users, management of communication between the ship and the shoreside, and development and implementation a cyber incident response plan based on a risk assessment.

American Bureau of Shipping (ABS) – ABS in 2016 released *The Application of Cybersafety Principles to Marine and Offshore Operations,* which provides best practices for cybersecurity across the marine and offshore communities to provide insights for operations, maintenance and support of cyber-enabled systems.[38] They have been updated to align with the *ABS Guide for Cybersecurity Implementation for the Marine and Offshore Operations,* which provides safety and security requirements and recommendations on the assessment of cybersecurity systems.[39] It also provides guidance for vessel readiness for preventing and managing cyber events that may compromise the safety and security of the data, systems, and vessels of a company or organization.

Digital Container Shipping Association (DCSA) – DCSA has published the *DCSA Implementation Guide for Cyber Security on Vessels* for the global container shipping industry.[40] The guide outlines best practices to assist company cyber security personnel who are responsible for fleet-wide cyber security onboard vessels. It provides guidance on cyber-related aspects of IT such as computers, electronic manuals and networks and applications OT that includes engine control, ECDIS, onboard measurement and control systems, PLCs and remote support for engines.

International Association of Classification Societies (IACS) – IACS has released its *Recommendation on Cyber Resilience* to help achieve flexibility in cyber security practices.[41] Areas of focus include the design, integration and/or maintenance of computer-based systems to support secure operations. Also discussed is protection against unauthorized access, misuse, modification, destruction, or improper disclosure of information created, archived, or used in onboard computer-based systems or transported through networks connecting such systems.

International Association of Ports and Harbors (IAPH) – In 2021, IAPH published its *Cybersecurity Guidelines for Ports and Port Facilities* to provide the international port industry with a set of cybersecurity guidelines based on successes achieved by ports and port facilities from around the world.[42] They are designed to assist executives in the port industry to foster greater collaboration within their organizations, as well as more broadly with their local, regional, national and international partners and stakeholders.

United States Coast Guard – The U.S. Coast Guard's (CGCYBER) Maritime Cyber Readiness Branch (MCRB) supports the cybersecurity mission in the commercial maritime transportation community.[43] They provide many resources that can assist in developing, maintaining and updating strategic planning and cyber policy; assessing threats, vulnerabilities and impact of loss to the Maritime Transportation System (MTS), sustaining strong partnerships with key MTS stakeholders to develop insight into future MTS threat vectors; and oversee cybersecurity support to MTS operations through threat assessments, cyberinfrastructure posture improvements, info sharing and cyber domain incident awareness.

National Institute of Standards and Technology (**NIST**) – The NIST released its *Framework for Improving Critical Infrastructure Cybersecurity*, focusing on using business drivers to guide cybersecurity activities and considering cybersecurity risks as part of the organization's risk management processes.[44] Released in 2018, the Framework consists of three parts: the Framework Core, the Implementation Tiers and the Framework Profiles. The Framework Core is a set of cybersecurity activities, outcomes and informative references that are common across sectors and critical infrastructure. Elements of the Core provide detailed guidance for developing individual organizational Profiles. Through the use of Profiles, the Framework will help an organization to align and prioritize its cybersecurity activities with its business/mission requirements, risk tolerances and resources. The Tiers provide a mechanism for organizations to view and understand the characteristics of their approach to managing cybersecurity risk, which will help in prioritizing and achieving cybersecurity objectives.

9.7 Future Directions to Enhance Cyber Security

Opportunities exist to create multiple solutions that enable rapid development and dynamic updates of accurate and comprehensive reduced-order models of the protect surface. This is especially important when dealing with complex critical infrastructure to detect critical system elements that would make them susceptible to model-based attacks. Novel artificial intelligence (AI)-based and other methods are needed to autonomously build and update these models to represent classes of network system functionalities that span multiple, different ship and port infrastructure configurations and defend against a multitude of diverse threat vectors. Such methods must minimize human-guided domain knowledge and training requirements and feature dynamic development, adaptation and reconfiguration to overcome vulnerabilities exposed by destabilization attacks. These must also result in an order-of-magnitude reduction in development time over existing methods, enhanced resilience to broad-based attacks and reduced brittleness to previously unknown cyber tactics.

Strategies and solutions to cyber security issues and problems in the United States are investigated by several agencies under the Executive Office of the President, several of which have already been identified in this chapter. The National Science and Technology Council (NSTC) is the principal means by which the Executive Branch coordinates science and technology policy across the diverse entities that make up the Federal research and development enterprise. Its primary objective is to ensure that science and technology policy decisions and programs are consistent with the President's stated goals. The NSTC prepares research and development strategies that are coordinated across Federal agencies aimed at accomplishing multiple national goals.

The 2019 *Federal Cybersecurity Research and Development Strategic Plan* aims to coordinate and guide U.S. Federally funded R&D in cybersecurity, including the development of consensus-based standards and best practices.[45] The Plan identifies four interrelated defensive capabilities (deter, protect, detect and respond) and six priority areas for cybersecurity R&D (artificial intelligence, quantum information science, trustworthy distributed digital infrastructure, privacy, secure hardware and software, and education and workforce development) as the focusing structure for Federal cybersecurity R&D activities and investments to benefit the Nation.

Signed into law in December 2022, the *Quantum Computing Cybersecurity Preparedness Act* represents major legislation in the area of cybersecurity that encourages the U.S. Federal government to adopt technology that is protected from decryption by quantum computing.[46] This act addresses the migration of executive agency information technology systems to post-quantum cryptography, where encryption is strong enough to resist attacks from quantum computers developed in the future. After the National Institutes of Standards and Technology (NIST) has issued post-quantum cryptography standards, guidance will be issued requiring each executive agency to develop a plan to migrate the information technology of the agency to post-quantum cryptography. Also included will be a strategy to address the risk posed by IT vulnerabilities to weakened encryption due to the potential and possible capability of a quantum computer to breach such encryption and the development of standards for post-quantum cryptography. Further information on quantum sensing is described in greater detail in Chapter 11, Next Generation Sensing.

References

1 Vanguard. Maritime Cyber Attacks Increase by 900% in Three Years. 29 July 2020. Vanguard Media Limited, Nigeria. www.vanguardngr.com/2020/07/maritime-cyber-attacks-increase-by-900-in-three-years/

2 MarineLink. Cyber Attacks on the Rise at US Ports and Terminals. 5 October 2022. www.marinelink.com/news/cyber-attacks-rise-us-ports-terminals-499964

3 Jones Walker. 2022 Ports and Terminals Cybersecurity Survey. Jones Walker LLP. www.joneswalker.com

4 Alejandro Mayorkas, Secretary of the U.S. Department of Homeland Security. Testimony before the U.S. Senate Committee Hearing Channels on Threats to the Homeland. 17 November 2022.

5 David Kushner. The real story of stuxnet. *IEEE Spectrum*. 26 February 2013. https://spectrum.ieee.org/the-real-story-of-stuxnet

6 Deborah Haynes. Iran's Secret Cyber Files. *Sky News*. 27 July 2021. https://news.sky.com/story/irans-secret-cyber-files-on-how-cargo-ships-and-petrol-stations-could-be-attacked-12364871

7 Kuo-Hui Yeh, Wenli Shang, Tianyu Gong, Chunyu Chen, Jing Hou, and Peng Zeng. (2019). Information security risk assessment method for ship control system based on fuzzy sets and attack trees. Security and Communication Networks, Article ID 3574675. https://doi.org/10.1155/2019/3574675

8 G. Kavallieratos, S. Katsikas, and V. Gkioulos. (2019). Cyber-attacks against the autonomous ship. In: Katsikas, S., et al. Computer Security. SECPRE CyberICPS 2018. Lecture Notes in Computer Science, vol. 11387. Springer, Cham. https://doi.org/10.1007/978-3-030-12786-2_2

9 World Maritime News. Nightmare Scenario: Ship Critical Systems Easy Target for Hackers, December 2017. https://worldmaritimenews.com/archives/238869/nightmare-scenario-ship-critical-systems-easy-target-for-hackers

10 B. Svilicic, J. Kamahara, M. Rooks, and T. Yano. Maritime cyber risk management: An experimental ship assessment. *Journal of Navigation*, 72(5), 1108–1120.

11 Offshore Energy. Nightmare Scenario: Ship Critical Systems Easy Target for Hackers. 21 December 2017. www.offshore-energy.biz/nightmare-scenario-ship-critical-systems-easy-target-for-hackers/

12 Bartlomiej Hyra. M.Sc. Thesis, Master of Science in Engineering. DTU Compute Department of Applied Mathematics and Computer Science, Technical University of Denmark, Kongens Lyngby (2019). https://backend.orbit.dtu.dk/ws/portalfiles/portal/174011206/190401_Analyzing_the_Attack_Surface_of_Ships.pdf

13 University of Texas. UT Austin Researchers Successfully Spoof an $80 million Yacht at Sea. 29 July 2013. https://news.utexas.edu/2013/07/29/ut-austin-researchers-successfully-spoof-an-80-million-yacht-at-sea/

14 Defense Aviation. Iran Displays Captured US RQ-170 Sentinel, 2011. www.defenceaviation.com/iran-displays-captured-us-rq-170-sentinel/

15 Corey D. Ranslem. Secure at Sea: AIS Fraught with Vulnerabilities. 10 January 2020. www.the-triton.com/2020/01/secure-at-sea-ais-fraught-with-vulnerabilities/

16 BBC News. Warship Positions Faked Including UK Aircraft Carrier. 2 August 2021. www.bbc.com/news/ technology-58027363

17 Eduard Kovacs. Ship Data Recorders Vulnerable to Hacker Attacks. Security Week. 11 December 2015. www.securityweek.com/ship-data-recorders-vulnerable-hacker-attacks/

18 Sean Gallagher. Hacked at Sea: Researchers Find Ships' Data Recorders Vulnerable to Attack. ARS Technica, 10 December 2015. https://arstechnica.com/information-technology/2015/12/hacked-at-sea-researchers-find-ships-data-recorders-vulnerable-to-attack/

19 Ruben Santamarta. Maritime Security: Hacking into a Voyage Data Recorder (VDR). IOActive. 9 December 2015. https://ioactive.com/maritime-security-hacking-into-a-voyage-data-recorder-vdr/

20 Danny Palmer. Critical IoT Security Camera Vulnerability Allows Attackers to Remotely Watch Live Video – and Gain Access to Networks. 17 August 2021. www.zdnet.com/article/critical-iot-security-camera-vulnerability-allows-attackers-to-remotely-watch-live-video-and-gain-access-to-networks/

21 Abhishek Bhanawat . What Are Intrinsically Safe Cameras Used on Ships? *Marine Insight*. 18 March 2019. www.marineinsight.com/marine-safety/what-are-intrinsically-safe-cameras-used-on-ships/

22 Thorsten Stremlau. The Vulnerability of Satellite Communications. *Security Magazine*. 19 April 2021. www.securitymagazine.com/articles/94689-the-vulnerability-of-satellite-communications

23 Jennifer Hlad. To Protect Satellites, Secure Your Networks, Chief of Space Ops Says. Defense One. 31 January 2023. www.defenseone.com/threats/2023/01/protect-satellites-secure-your-networks-chief-space-ops-says/382424/

24 Jessie Hamill-Stewart and Andrew Sallay. VSAT Connectivity Comes with Cybersecurity Threats to Vessels. Maritime Executive. 23 May 2023. www.maritime-executive.com/editorials/vsat-connectivity-comes-with-cybersecurity-threats-to-vessels

25 John Harries. A Fundamental Danger of NMEA 2000 Networks. 26 May 2013 and updated 11 July 2021. www.morganscloud.com/2013/05/26/nmea-2000 missing-the-obvious/

26 Trend Micro, Incorporated. Cyberattacks against Intelligent Transportation Systems: Assessing Future Threats to ITS, 2017, pp. 27–28. https://documents.trendmicro.com/assets/white_papers/wp-cyberattacks-against-intelligent-transportation-systems.pdf

27 Zeinab El-Rewini, Karthikeyan Sadatsharan, Niroop Sugunaraj, Daisy Flora Selvaraj, Siby Jose Plathottam, and Prakash Ranganathan. 22 June 2020. *IEEE Sensors Journal*, 20(22), ISSN 1530-437X.

28 Vincenzo Ciancaglini, Joey Costoya, Philippe Lin, and Roel Reyes. Exploitable Flaws Found in Facial Recognition Devices. *Trend Micro*. 22 September 2020. www.trendmicro.com/en_us/research/20/i/exploitable-flaws-found-in-facial-recognition-devices.html

29 Safetytech Accelerator. Startup Deploys Wireless Sensors across Ship's Cargo Hold to Predict Fire. 3 March 2020. https://safetytechaccelerator.org/case-studies/safetytech-startup-deploys-wireless-sensors-across-ships-cargo-hold-to-predict-fire/

30 Jasper Helder and Lars-Erik Hjelm. Focus on shipping in fight against drug smuggling. *Lloyds List*, 1 May 2020. https://lloydslist.maritimeintelligence.informa.com/LL1132202/Focus-on-shipping-in-fight-against-drug-smuggling

31 Katie Terrell Hanna. Attack Surface Definition. *WhatIs.com*. September 2021. www.techtarget.com/whatis/definition/attack-surface

32 U.S. Department of Defense. Department of Defense Releases Zero Trust Strategy and Roadmap, 22 November 2022. www.defense.gov/News/Releases/Release/Article/3225919/department-of-defense-releases-zero-trust-strategy-and-roadmap/

33 John Kindervag. Define a Protect Surface to Massively Reduce Your Attack Surface. *Palo Alto Networks*, 4 September 2018. www.paloaltonetworks.com/blog/2018/09/define-protect-surface-massively-reduce-attack-surface/

34 Some of this information was obtained from: www.portsamerica.com/locations/portofbaltimore-maryland

35 IMO MSC-FAL.1. Guidelines on Maritime Cyber Risk. MSC-FAL.1/Circ.3/Rev.1, 14 June 2021. wwwcdn.imo.org/localresources/en/OurWork/Facilitation/Facilitation/MSC-FAL.1-Circ.3-Rev.1.pdf
36 IMO MSC.428(98). Resolution MSC.428(98). Maritime Cyber Risk Management in Safety Management Systems, 16 June 2017. wwwcdn.imo.org/localresources/en/KnowledgeCentre/IndexofIMOResolutions/MSCResolutions/MSC.428(98).pdf
37 BIMCO, et al. The Guidelines on Cyber Security Onboard Ships. Baltic and International Maritime Council. Version 4, 2020. www.bimco.org/about-us-and-our-members/publications/the-guidelines-on-cyber-security-onboard-ships
38 ABS Cybersafety Vol. 1. Guidance Notes on The Application of Cybersafety Principles to Marine and Offshore Operations. September 2016. Vol. 1. American Bureau of Shipping. Spring, TX, USA. ww2.eagle.org/content/dam/eagle/rules-and-uides/current/other/250_cybersafetyV1/CyberSafety_V1_Cybersecurity_GN_e.pdf
39 ABS Cybersafety Vol. 2. ABS Guide for Cybersecurity Implementation for the Marine and Offshore Operations. February 2021. Vol. 2. American Bureau of Shipping. Spring, TX, USA. ww2.eagle.org/content/dam/eagle/rules-and-guides/current/other/251_cybersafety_2021/cybersafety-v2-cybersecurity-guide-feb21.pdf
40 DCSA. DCSA Implementation Guide for Cyber Security on Vessels v1.0. Digital Container Shipping Association. 3 October 2020. https://dcsa.org/wp-content/uploads/2020/03/DCSA-Implementation-Guideline-for-BIMCO-Compliant-Cyber-Security-on-Vessels-v1.0.pdf
41 IACS. Recommendation on Cyber Resilience. International Association of Classification Societies. No. 166. IACS Rec. 2020/Corr.2. 2022. www.iacs.org.uk/news/iacs-launches-single-standalone-recommendation-on-cyber-resilience/
42 IAPH Cybersecurity Guidelines. IAPH Cybersecurity Guidelines for Ports and Port Facilities. International Association of Ports and Harbors, Tokyo, Japan. Version 1.0. 2 July 2021. https://sustainableworldports.org/wp-content/uploads/IAPH-Cybersecurity-Guidelines-version-1_0.pdf
43 USCG MCRB. U.S. Coast Guard Maritime Cyber Readiness Branch (MCRB), Alexandria, VA. www.dco.uscg.mil/Our-Organization/CGCYBER/Maritime-Cyber-Readiness-Branch/
44 NIST. Framework for Improving Critical Infrastructure Cybersecurity. Version 1.1 National Institute of Standards and Technology, 16 April 2018. https://nvlpubs.nist.gov/nistpubs/CSWP/NIST.CSWP.04162018.pdf
45 NSTC. Federal Cybersecurity Research and Development Strategic Plan. National Science and Technology Council. December 2019. www.nitrd.gov/pubs/Federal-Cybersecurity-RD-Strategic-Plan-2019.pdf
46 Quantum Computing Cybersecurity Preparedness Act. Public Law117-260. www.congress.gov/bill/117th-congress/house-bill/7535

Chapter 10

Vessel Sensor Degradation

The most reliable means for detecting sensor degradation has traditionally been well-trained mariners with years of experience who can readily distinguish between normal operation and anomalies in sensor performance throughout a wide range of working scenarios. Unfortunately, there is a shortage of well-trained mariners that is not expected to be resolved soon making this problem more significant.[1] As ship automation continues to increase and sensor systems gain added complexity, the detection of sensor degradation becomes even more difficult. Further complicating matters is the introduction of unmanned and autonomous vehicles and their need for enhanced and extended sensor suites and sensor data fusion technology that allows them to operate without seafarers onboard. Still, the problem of sensor degradation detection remains unresolved.

Unmanned and autonomous vehicles function as passenger, cargo and other types of vessels included in the partial list cited in the first chapter of this book. These same technologies continue to be integrated on the bridges of conventionally staffed vessels to assist seafarers in the performance of their duties. The volume of imagery, signal and digital data generated by sensors in recent years has increased by orders of magnitude to advance the development of groundbreaking hardware and software leading to new and improved processes that make better use of these resources. The promises of innovations in system design and function have only begun to be tapped and include providing new perspectives from which enhanced situational awareness and safety of navigation may be achieved.

Much research has been performed, and entire industries established to account for and deal with the failures of sensors, the various failure modes to which they are susceptible and the many methods and techniques that can be applied to prevent, detect and even predict failure. However, little has been said and even less investigated regarding the insidious degradation of sensors especially over extended periods of time that can cause misleading and deceptive indications and shield anomalies. Such occurrences can stymie reliable and dependable ship operations and reduce overall safety.

DOI: 10.1201/9781003377900-10

Degradation implies one or more aspects of sensor data corruption that can reduce the quality of the service they provide even though the sensor itself may be fully operational and pass all functional test parameters. Obvious examples of sensor degradation include:

- Mud or other contaminant on the lens of a camera that partially blocks its view.
- A close-by obstruction in the partial field of view of a Radar antenna that causes a dead zone without returns on a Radar display.
- Excessive barnacle and other growth on a Sonar transducer that reduces effective range.
- Interference from other systems causing strange behavior in a display.
- Corrosion in a transmission line resulting in excessive impedance that can produce unexpected results.

Sensor degradation caused by human agencies with nefarious intent includes physical and cyberattacks that are also widespread and problematic. Many such events can often be diagnosed by a mariner who has gained a high level of expertise using such systems. However, watchstanders with learned knowledge of sensor theory but little practical experience in sensor operation will not necessarily be capable of reliably discerning degradation. More importantly, unmanned vessels and autonomous vehicles have even less ability to do so.

Less obvious examples of sensor degradation include the gradual loss of pixels in a camera over time that steadily reduces resolution and accuracy, the effects of precipitation and fog on light propagation and infrared night vision performance, and inadequately documented performance anomalies in new types of sensors recently introduced onto the bridge. Similar effects can be identified in engineering applications such as optical devices that become contaminated and grow cloudy, proximity sensors that shift from optimal alignment and resistive transducers that age prematurely and exceed specifications due to environmental and other factors.

This chapter highlights much of the original research performed by the author and others aimed at detecting and overcoming sensor degradation. Applications for this technology span many industries, much of which has focused on the automotive and trucking industries. However, comprehensive cases are made for sensor degradation detection technology use in the maritime sector to enable unmanned surface and undersea vehicles to operate much in the same fashion as a conventionally staffed vessel, and to assist the navigation and engineering watches on conventional ships in the performance of their duties. What follows are illustrations of the fruits of these technologies when applied to solve real-life problems and application scenarios aboard ships and maritime vehicles.

10.1 Background and Investigative Approach

The ability to detect and recognize both rapid and subtle degradation in sensor performance over time and, if possible, compensate for their effects is essential to continue and maintain the nominal performance of ship operations and vehicle systems in a safe manner. This represents a highly significant limitation that stymies commercial and naval vessels as well as unmanned maritime vehicle operations. Also significant is the means to communicate sensor status and degradation characteristics to one or more operations centers, vessel traffic services and Remote Control Centers (RCCs); and other ships and maritime vehicles via broadband and low-bandwidth data links to collaborate and to provide periodic system health status. Such capabilities enhance ship and vehicle resilience to achieve voyage objectives and provide a means to continuously learn new tactics to defeat or minimize such threats. Efforts by several researchers have resulted in various methods to detect sensor degradation as well as internal sensor failures. They have also resulted in new ways to make systematic adjustments necessary to mitigate and compensate for reduced and degraded, corrupted and perplexing sensor inputs to increase the probability of a successful voyage.

One approach involves the performance of sensor signal analytics on a variety of navigation instruments through sensor data correlation from three different perspectives.[2] These include the distribution of signal characteristics across the frequency spectrum at any point in time (Frequency Domain), changes that take place in the signal over time (Time Domain) and, where applicable, processed sensor signals as represented in imagery (spatial domain) created from Frequency and Time Domain signal data.[3] Machine learning and deep-learning artificial intelligence (AI) are used to detect characteristic signatures and trends in sensor data indicative of rapid as well as gradual sensor degradation from various natural and manmade phenomena. Detection of degradation events consider failure modes and effects due to external causes such as grounding and allision, extreme sea states and physical damage; and internal causes resulting from failure in vehicle propulsion, steering, electrical, hydraulic, communications, navigation and health monitoring systems. This three-domain approach facilitates the classification of a broad range of sensor types and signal characteristics and provides a basis for the detection of sensor degradation, the prognosis of pending and actual sensor and system failures, and the implementation of alternative mitigation and recovery methods tailored to the unique attributes of each cause.

A second approach explores the effects of ultrasound noise on the performance of navigation instruments utilizing a gyroscope, accelerometer, and magnetometer.[4] Considered in this approach is sensor data fusion from a number of sensors to produce more accurate information than if data were taken from individual sensors and used separately. The primary goal is to enhance performance and reliability by detecting inconsistencies in sensor measurements when under the influence of acoustic and magnetic emissions external to the system.

A third approach focuses on the influence of electromagnetic interference on Ethernet networks on critical sensor systems.[5] Specifically considered is their effect on Time-Triggered Ethernet (TTE) in terms of precise signal timing, redundant pathways and careful switching to ensure that information arrives precisely when needed. TTE is already in use in flight-critical aircraft systems, aboard spacecraft and power-generating wind turbines and its use is becoming more prevalent on ships and maritime vehicles.

This chapter discusses a group of sensors most applicable and common to ships and maritime vehicles selected and weighted by priority for voyage requirements as well as safety of navigation. These sensors include Sonar, Radar, visual and infrared cameras, Lidar, and other systems through which vehicle situational awareness related to external factors, events and threats is maintained. Also considered were internal vessel sensors, such as exist for a wide variety of engineering system sensor applications, as well as inertial navigation (INS) and other systems that require periodic updates of position, navigation and timing (PNT) inputs to maintain accuracy by minimizing their susceptibility to increasing progressive error over time.

Decision-making capabilities were also developed using machine learning and deep-learning AI techniques to determine potential sources (sensors and sensor inputs), types (rapid or gradual occurrence) and symptoms (noise, distortion, etc.) of sensor degradation. Considered is whether sensor degradation is very likely to occur soon or is currently in evidence and the various alternatives to mitigate and compensate for the effects of degradation. These alternatives include solutions targeted to individual sensor systems as well as approaches that consider the fusion of alternative and secondary sensor data sources to derive an acceptable solution. Theoretical assumptions regarding both the causes of sensor degradation and their effects were validated through experimentation using a research vessel at sea in an operational environment for evaluating new technologies for civilian ships and unmanned, remotely operated and autonomous surface vehicles. Inherent to this solution is the ability to perform operations in communications-denied environments without assistance from remote controllers or observers or with limited communications where small data packets containing relevant sensor information can be sent to provide status and diagnostic information to a remote location. The approach followed examines a constant stream of external and internal vehicle sensor data, continually assesses data characteristics for evidence of potential and actual degradation and, upon detection, potentially recommends and implements alternative courses of action to maintain readiness, preserve the safety of navigation, and achieve voyage objectives.

10.2 Sensor Systems Examined

A comprehensive suite of navigation and engineering sensors and sensor types with a variety of inputs, outputs and signal characteristics were targeted in this research. External causes of sensor degradation originating from outside the ship or vehicle were considered, along with causes originating from within the vessel due to internal system performance outside of nominal limits (meaning degraded) and failure that may cause direct and unintended indirect consequences. This includes both onboard and offboard sensors and related systems that provide situational awareness of relevant phenomena from the perspectives of space, at the water's surface and below the waterline. More than 24 systems were considered, including Sonar, Radar, Lidar, visible light and infrared imaging, GNSS, meteorological, oceanographic and other types of sensors. Also considered were internal vehicle system sensors associated with propulsion, steering, electrical, communications, and navigation, where both rapid and gradual sensor degradation can occur. This includes INS, where the absence of external references for PNT data results in progressively increasing errors in velocity that are compounded into even larger errors in position resulting from integration drift. Prognostics of imminent failure and diagnosis of degraded performance in vehicle engineering systems can also be detected using noncontact sensors. Examples include electronic signatures of individual motors, pumps and other individual equipment indicative of worn relays, burned windings, corrosion, broken insulation and an increasing lack of efficiency for specific system components.[6]

A subset of these was selected for experimentation, including five types of Sonar (echosounder, multibeam, side-scan, chirp and navigation) and two types of imaging (visible light and infrared) sensors; with additional consideration being given to Radar and Lidar. Internal vehicle sensors include INS and sensors associated with alternators, tanks and other onboard systems. These selections were based in part on applicability across a broad range of ship and vehicle types, diversity of data and signal types, and access to intermediate data, signal elements and endpoint imagery representative of common sensor network inputs. Analysis of signal characteristics is performed as they appear in two or three domains, as applicable, during nominal operation and while experiencing both rapid and gradual performance degradation over longer periods of time. Causalities of sensor degradation were considered including naturally occurring environmental conditions, interference from internal and external sources, and damage from hazards to navigation, adversarial action and internal sensor failure modes.

10.3 Sources of Sensor Degradation

External causes of sensor degradation considered included marine growth on hull exteriors and sensor transducers; interference, electrical and physical damage to sensors caused by grounding and allision, as well as the effects of nefarious acts by adversaries. Internal sensor system failure modes were considered where their effects can be ambiguous and hard to distinguish yet are vital to the development of an effective mitigation and compensation strategy. Sensor degradation is considered on the basis of the symptoms of its occurrence and their effects as may be attributed to changes in signal characteristics from both external causes and causes internal to the vehicle itself. This is accomplished through system failure and is based upon findings in the relevant literature as well as experimental results related to failure prognostics, fault diagnosis and sensor analytics. However, practical aspects of software design and development, training data set creation and the development of cases for verification, validation, and testing require the acquisition and use of very large volumes of data and imagery. These datasets must be obtained in a relevant operational environment to train machine learning and deep-learning AI solutions to adequately perform these tasks. Different types of experiments were performed that provided direct evidence of sensor degradation over time. Training data sets were created consisting of millions of instances of discrete signal measurements, characteristics, imagery and metadata for nominal and aberrant sensor performance at increasing degrees of hull fouling, interference, damage and internal failures. Experiment goals and methods are specific to each undertaking and described in the paragraphs that follow.

10.3.1 Biofouling

Gradual sensor degradation over time by decreasing sensor efficiency is caused by biofouling, which is a result of the growth of marine plant and animal life on underwater transducers. This growth can attenuate Sonar and other sensor signals transmitted from active sensors as well as reflected and passive signals received by onboard sensors. In addition, hull roughness attributed to biofouling greatly reduces hull efficiency resulting in reduced speed given the same engine thrust, greater wear and tear on propulsion and steering systems, and differences in vehicle handling and control surface response that can alter variables used to predict INS progressive error. This is particularly critical for unmanned vehicles expected to be deployed at sea for months at a time without manned intervention. During such long deployments significant fouling would be expected to occur. It is necessary to ascertain the effects of fouling on sensor signals through direct measurement in the Frequency and Time Domains at the sensor and transducer interface and, where applicable, determine the symptoms of fouling on Spatial Domain representations of processed transducer signals.

10.3.2 Interference

Numerous different experiments took place over a span of many months to simulate and effect actual rapid and gradual sensor degradation in a relevant operational environment due to interference from natural and manmade sources both innate to the natural environment and resulting from adversarial action. This includes spoofing and denial of service (DoS) attacks on GPS/GNSS systems themselves, and sensor systems that depend on continuously updated PNT data. Also included in these experiments was the use of multiple extraneous Radar, Sonar and Lidar signals directed toward normally operating systems with the intention of jamming, confusing and degrading these systems outside of normal operating specifications.

Degradation of visual and infrared sensors was also accomplished using actual environmental conditions known to facilitate degraded performance that includes various precipitation intensities (light, moderate and heavy) and types (rain, snow, sleet, hail, etc.). Experiments were performed under varying visibility from clear and unlimited conditions to zero visibility due to fog and other causes. The effects of icing, turbulence, condensation, and other factors, such as a buildup of salt deposits from sea spray that degrade sensor performance were also considered through natural occurrence and/or simulation of these conditions. A simulation of GPS/GNSS interference with INS was performed to eliminate any potential for interference with the navigation of nearby aircraft and vessels.

10.3.3 Damage

Many ways exist to damage sensors in a manner that falls short of complete failure and results in degraded capability. For external sensors, these include misalignment and partial blockage of view from being hit with debris as well as from large wave strikes, groundings and allision that causes physical damage to sensors and transducers. Also considered is the buildup of contaminants from the combustion of nearby vessel components, intense heat from fire or internal overheating caused by lack of ventilation, extreme cold from severe operational conditions, glancing strikes by laser weapons, and even the application of chemicals designed to render sensors blind or otherwise inoperable. For internal sensors, performance degradation can originate from failures within the various propulsion, electrical, hydraulic, communication, navigation, and other systems within the vehicle. A subset of vehicle temperature, pressure, voltage, current, power and other sensors and measurement characteristics commonly associated with these systems were selected for demonstration.

10.3.4 Degradation Modes

Experiments associated with sensor degradation differ from the sensor damage experiments described in the previous paragraph as the causes of both rapid and gradual degradation can be due to anomalies integral to the technologies used and the design of the system components themselves, and not necessarily caused by external factors. Examples of visual and infrared cameras include the presence of many horizontal lines across an image as well as splashes of color that represent a bad sensor element rather than external damage. The causes of such failures can be difficult to discern and range from electrostatic discharge to laser strikes, excessive heat, humidity, loose electromechanical connections and long-term exposure to sunlight.

Sonar degradation modes that can cause data corruption include loss of transducer watertight integrity, electrical shorts, electrical arcing on the transducer, low insulation resistance and thermal runaway. Additional failure modes exist with respect to the connectors themselves. These anomalies can manifest as noise within sensor signal measurements, spots and pixilation across images, lack of dynamic range, partial or full image freezing, and many more symptoms peculiar to the types of sensors used.

These and other degradation modes have been identified for each of the sensor types considered. Failure modes were simulated or physically introduced into a number of these systems to measure sensor performance and signal characteristics across the two or three domains, as applicable. Measurement datasets representative of sensor degradation modes were also created for use in developing and training machine learning and deep-learning AI sensor capabilities.

10.3.5 Symptoms and Characteristics

Signatures of sensor degradation detected within electrical signal characteristics across the frequency spectrum can include reduced or excessive signal strength in transmitted and received signals; the presence of ringing, inappropriate signals or undesirable harmonics, the absence of required signals and incorrect signal characteristics. Change in signal characteristics over time can provide a direct correlation to causative events and insight into rates of sensor degradation, provide context regarding the nature and severity of the anomaly, and possible alternative courses of corrective action. For more complex sensors such as Sonar, Radar, Lidar, and visual/infrared devices where signal processing of transducer data results in the creation of a graphic representation, degradation can be manifest as noise, distortion, pixelation, incompleteness, loss of resolution, and other as Spatial (or pixel) Domain anomalies within the imagery. For INS, we examine the fusion of sensor inputs used to update PNT and consider the potential effects of GPS/GNSS signal degradation and unavailability, spoofing, and DoS attacks.

10.4 Sensor Degradation Detection

Methods used and processes followed are described for the development of capabilities to detect, identify and mitigate sensor degradation resulting from various causalities. This includes, were possible, prognosis and diagnosis of sensor system failures to distinguish them from external causes of sensor degradation. Details of the characteristics of system components are provided relevant to sensor degradation detection, the development processes used and the methods and procedures followed in evaluating the results.

10.4.1 Design and Implementation

The design of system hardware components, software architecture, machine learning, and deep-learning AI to implement the prognostic and diagnostic elements of sensor degradation was accomplished based on the specific requirements associated with the selected sensor types. Also included were capabilities for mitigation and compensation for sensor degradation.

System hardware is comprised of commercial-of-the-shelf (COTS) sensor components providing digital, analog to digital conversion; NMEA, Ethernet and other data bus capabilities, and a COTS computer. A software-defined radio (SDR) performs oscilloscope and spectrum analyzer functions needed to accomplish sensor data analytics for Frequency and Time Domain signal measurements and characteristics. Special data conversions are also performed as may be needed to support the unique requirements of vehicle sensors, such as resistance, voltage, current and other measurements.

The computer controlled SDR operations and received SDR sensor data analytics products for further analysis and correlation between the Frequency and Time Domains. It also performed Spatial Domain analytics using imagery products created by more complex sensors such as Sonar, Radar and Lidar. Embedded within the computer are sensor data reasoning capabilities in the form of machine learning and deep-learning AI solutions responsible for analyzing three domain data. This determined characteristics needed to determine whether sensor degradation exists now or may exist in the immediate future, and identified specific details regarding sensor degradation. In addition to external sensors, the computer also acts as the primary interface to the vehicle's internal propulsion, steering, electrical, hydraulic, communications, navigation, and health monitoring systems.

10.4.2 Rapid Prototyping

Rapid prototyping helped to quickly create a working software architecture and functional implementation to implement effective reasoning capabilities for one type or class of sensors and sensor signal characteristics. Subsequently, the system's scope was widened to encompass additional sensor systems and characteristics. Each sensor type represents one iteration of an incremental development model where design, development and testing were accomplished for the set of new functions introduced during each cycle. In addition, retesting of the results of previous cycles was performed to ensure these capabilities still function properly and have not been adversely affected by newly developed capabilities. This process continued until an engineering model of the system was created and made available for testing over a period of several months.

10.4.3 Verification, Validation and Test

Determining correctness throughout our development processes included final engineering model development prior to testing. Verification involved determining that sensor requirements were fulfilled and that they had been correctly implemented. This process identified any missing requirements that should be considered, unnecessary requirements, and that the design process effectively implemented the requirements. Performance validation was accomplished to ascertain that the system is capable of correctly implementing the requirements. The validation process determined whether the requirements were reasonable in terms of their ability to reflect the actual functions that the product must perform, and whether the design was adequate to achieve the desired functions. Performance testing covered the breadth and depth of system functionality to determine whether it is robust and capable of performing across all required conditions and operational settings. This included exercising all system functions with a wide variety of test cases to demonstrate its capability to perform required functions and to ensure that, in the event of failure to perform properly, it is able to do so in a manner that is not destructive and provides traceability as to the cause(s) of failure.

While it is possible to achieve verification and validation of machine learning solutions using statistical processes, deep-learning AI solutions are not amenable to this approach due to the inherent lack of visibility into the functioning of their internal processing elements. Merely implementing black box testing techniques based on system requirements and design is insufficient as decisions are being made in real time. It may not be possible to effectively reproduce the same outcome without replicating the same exact stimulus recorded by sensors expressed within Frequency, Timeand Spatial Domain representations. This is presently an area of active research being actively pursued through many academic and industry initiatives, and by the U.S. DoD under DARPA's Assured Autonomy project.[7] Results to date have focused primarily on visualizing neural network activity by using activation maximization that attempts to determine what weights of individual node of a neural network provide the greatest amount of activation. Another considers determining how often specific paths through a neural network are activated using a "heat map" approach to visualizing neural network activity. Other efforts are also presently underway that may or may not provide useful solutions to this problem. Our approach examined implementing deep-learning AI solutions to the best of our ability and tracking the progress of others in this area. However, attempting to achieve a definitive solution to the verification and validation of deep-learning AI was beyond the scope of this effort.

Table 10.1 Sensor Suite Matrix

Sensor	*Manufacturer*	*Model*
Visual Camera	GoPro/SCW	Hero8/Gladiator 8, Detective 4
Infrared Camera	FLIR	MD625
Radar	Furuno/Garmin	1953C/Fantome 24
Sonar	*five manufacturers*	*various models*
Lidar	Ouster	OS2-128
GPS/GNSS	Furuno/Garmin	GP-37/8616xsw
Inertial Navigation	Advanced Navigation	GNSS/INS
Heading Sensor	Garmin	GAX-1141720
Automatic Identification System	ICOM/Garmin	MA-500TR/AIS-800
Engineering	Maretron	DSM-570
Weather	Maretron	WSO100
Satellite	SiriusXM	Marine
Microphone	SCW	Detective 4
Electronic Navigation Chart	Garmin	BlueChartG3

10.4.4 Physical Research Environment

Development and testing were accomplished using many different engineering and navigation sensors, a partial list of which is shown below in Table 10.1.

These sensors were located onboard the 10-meter research vessel outfitted with sensors, as illustrated in Figure 10.1. Located approximately 24 feet above the waterline, the Lidar sensor is installed atop the navigation light array with clear visibility through 360° free of all physical obstructions. Atop the Radar mast is a Furuno 12 kw, 72 nm (analog) Radar with a 48-inch open array antenna and a Garmin 40 w, 48 nm digital Radar with a 24-inch radome. Various sensors are located on a platform below the Radar mast, including GPS and GNSS antennae, satellite receiver antenna and weather sensor.

Beneath this platform are mounted six low-light video cameras, each with a 60° field of view for a total 360° coverage with internal microphones. Also mounted below this platform are six additional cameras, each also with a 60° field of view for a total 360° coverage to supplement the other cameras in providing flexibility with capture areas and rates along with options for image stability. Mounted forward and above the pilothouse is a very high-resolution bullet camera aimed directly ahead of the vessel with a motorized varifocal lens for optical zoom providing a range of fields of view, with the FLIR infrared camera mounted directly alongside. The heading sensor is located below the pilothouse floor at the approximate center of the vessel.

Four Garmin Sonar systems are installed on the vessel providing echo-sounder, side scan and forward-looking navigation capabilities. The

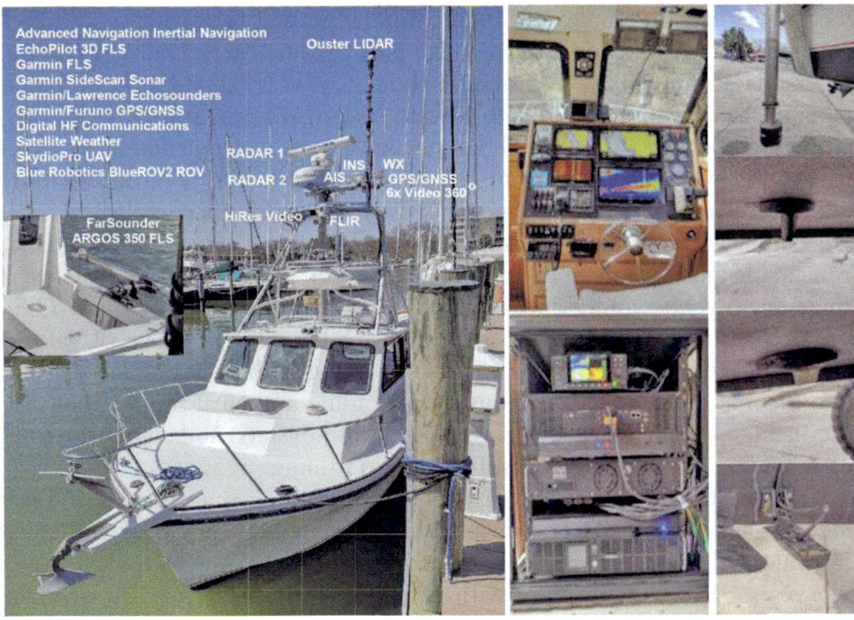

Figure 10.1 Ten-Meter Vessel Autonomy Testbed.

FarSounder forward-looking Sonar transducer is affixed to an over-the-side mount shown aft on the starboard side of the vessel behind the pilothouse.

A pair of EchoPilot forward-looking Sonar transducers are mounted athwartships approximately one-third of the vessel's length from the stern. Another echosounder transducer is located on the starboard side and the side-scan transducer is mounted on the stern of the vessel, with one additional echosounder transducer (not shown) mounted on the starboard side on the bottom near the stern. All Sonar transducers have been mounted at locations on the hull that ensure minimal or no interference will occur with other sensors.

Vessel sensors are routed internally via direct connection and Ethernet, NMEA-2000 and -0183 data bus to the helm console displays and controls. These signals are also routed to the instrument rack mounted in the forward cabin of the vessel that contains the computer system, sensor data recorders, sensor-specific computers, multi-display video monitor and uninterruptable power supply.

10.5 Detection Results

Results obtained across the areas studied are provided in the appropriate headings of this paragraph.

Please Note – Examples of nominal and degraded performance are illustrated using fully functional sensors that pass parametric and self-test criteria. Yet the data and display information they provide is tainted at best, and at worst misleading and unusable to where vessel operations and safety of navigation may be jeopardized. Such knowledge can be instrumental in reconfiguring onboard sensor systems to compensate for the effects of degradation, and to determine whether voyage objectives may still be accomplished.

Benchmark biofouling measurements of nominal signal characteristics were made on a clean hull using two different types of Sonar systems: a single-beam echosounder and a Forward-Looking Navigation Sonar (FLS). Signal characteristics were also measured while in the presence of various degrees of hull fouling, ranging from soft fouling with bacterial and diatomaceous slime and algae, calcareous macrofouling and hard fouling that may also include barnacles, tubeworms, mussels and oysters; and composite fouling including elements of both soft and hard fouling.

A testbed for this purpose was established using an echosounder and fish finder Sonar, and a 3D Forward-Looking Navigation Sonar. The testbed transducers and hull were continuously immersed in salt water with over two years of fouling in place with a fouling rating (FR) of 100 over the entire hull's wetted surface.[8] The EchoPilot navigation Sonar was fully functional but partially inoperative due to the extensive degree of fouling, and the Lowrance echosounder and fish finder was operational while exhibiting sensor degradation in terms of loss of range and resolution. Several iterations of cleaning of the hull took place to progressively reduce Fouling Rating from 100 to 0, with signal measurements taken at each degree of fouling. Although not a perfect replication of natural progression of fouling, this experiment provided us with rapid results over the course of several days that approximate naturally progressing fouling conditions. Experiment results were documented with signal characteristics and measurement values taken at each stage of fouling denoting electrical performance across the three domains for both testbeds, with Figure 10.7 (bottom) illustrating the effects.

The following figures provide examples of sensor degradation due to interference, damage and anomalies for the various sensors. Unless otherwise noted, each of these sensor degradation events were effectively and consistently detected.

10.5.1 Visible Light Camera

Camera images shown in Figure 10.2 depict various types of natural and manmade degradation that include fog, heavy precipitation, physical damage, chemical obscurants and physical obstructions. These are just a few of the many images obtained from the three types of onboard visual cameras listed in Table 10.1 under different weather conditions at different times of day that were used in testing image degradation detection capabilities. All

Figure 10.2 Degraded Visible Spectrum Camera Images.

illustrated states of degradation were detected in addition to many more that are not illustrated. The drone aircraft that operates off the research vessel also provides live video and data feeds amenable to the use of sensor degradation detection technology.

10.5.2 Infrared Camera

Figure 10.3 illustrates nominal (left) and degraded (right) imagery obtained from the infrared camera. Sensor degradation was accomplished as a result of heavy rain showers. The degree of degradation encountered was generally proportional to shower intensity.

Additional degradation was accomplished using many of the same techniques performed for visual camera degradation shown in Figure 10.2 with similar results. This includes the use of chemical obscurants, physical obstructions and weather conditions such as fog. All such degraded states were detected.

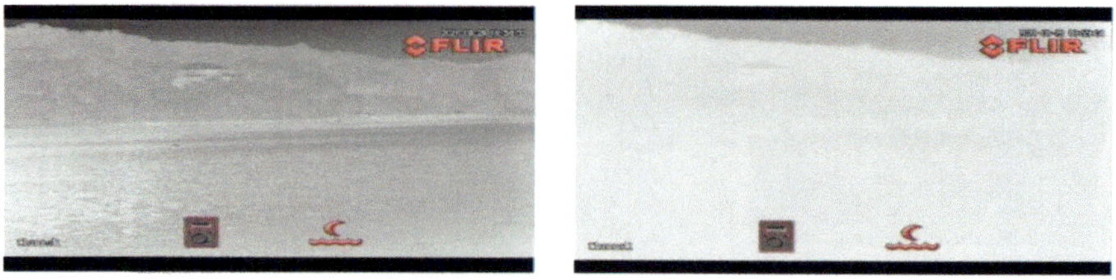

Figure 10.3 Nominal and Degraded Infrared Camera Images.

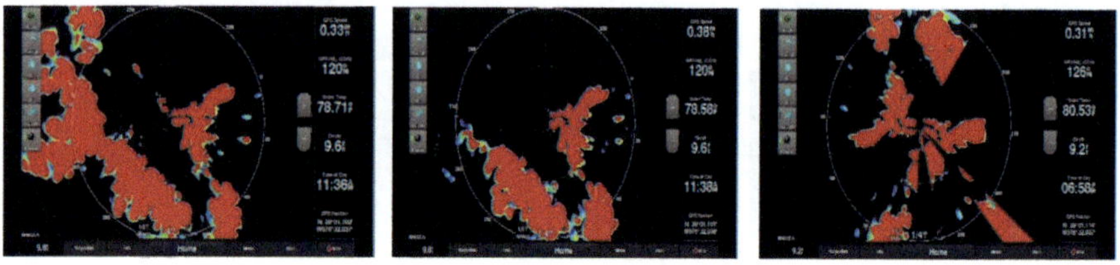

Figure 10.4 Nominal and Degraded Radar Images.

10.5.3 Radar

Figure 10.4 illustrates nominal (left) and degraded (center, right) images representative of different types of degradation that takes advantage of various combinations of events, timing and signal characteristics to facilitate detection. All images provided below are from the Garmin 40 w, 48 nm digital Radar system. The degraded image in the center illustrates the effect of an obstruction introduced into the signal propagation path. Small, partial obstructions through complete blockage of Radar signals was employed for these experiments with varying time durations.

The degraded image to the right depicts the result of electromagnetic interference with the heading sensor upon which the Radar system depends for proper geospatial orientation. The primary effect of this experiment was to cause the Radar display to gyrate wildly resulting in small to extreme distortion of the displayed imagery. Similar results were obtained with the Furuno 12 kw, 72 nm (analog) Radar using the same sensor degradation techniques. Both degraded states were detected in addition to several more that are not illustrated.

10.5.4 Sonar

Much of our effort focused on exploring the different types of degradation that may be experienced by the various types of Sonar sensors likely to be found on both conventional and unmanned vessels. These sensors included side-scan, traditional echosounders, forward-looking navigation and chirp Sonars produced by manufacturers that include Furuno, Garmin (shown here), Lowrance, FarSounder, and EchoPilot.

10.5.4.1 Side-Scan Sonar

Figure 10.5 illustrates side-scan Sonar images in both their nominal (top) and degraded (middle and bottom) states. Degraded states were accomplished through the use of physical obscuration over a portion of the sensor field of view, introduction of bubble screens and other natural phenomena into the water column and the physical removal of the transducer from the water as a result of performing high-speed turns with the vessel. All illustrated degraded states were detected in addition to several more that are not illustrated.

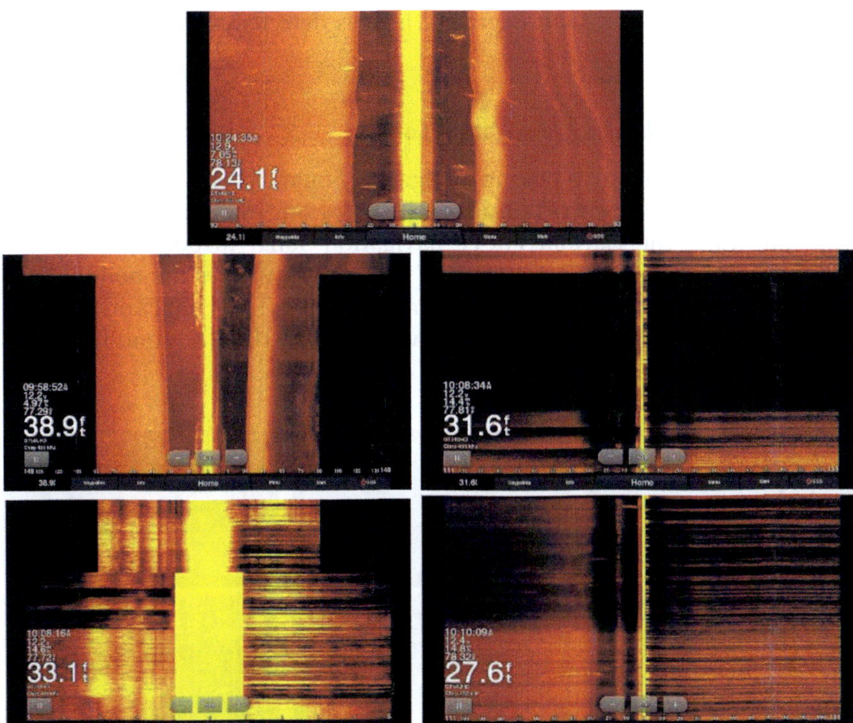

Figure 10.5 Nominal and Degraded Side-Scan Sonar Sensor Images.

10.5.4.2 Fish Finder Echosounder

Degradation experiments for echosounders with visual displays were performed where the image on the left of Figure 10.6 (top) obtained from the Garmin instrument depicts nominal echosounder performance while the center and right images show degraded performance. Note that some vessels are likely to be equipped with older and more primitive echosounders that merely provide a numerical indication of depth below the hull. Such devices are generally no-longer used in the commercial marketplace and the research vessel is not so equipped. However, our underwater remotely operated vehicle (ROV) has high-resolution depth/pressure and temperature sensors that are also used for such experiments.

10.5.4.3 Chirp Sonar

The scanning Sonar illustrated in Figure 10.6 (bottom) provides a nearly photographic, wide image below the vehicle where structure, submerged objects and fish can be discerned in nominal conditions. The nominal image is provided on the left, with degraded images detected shown in the center and right.

10.5.4.4 Forward-Looking Navigation Sonar

Several different forward-looking navigation Sonar systems were examined, and their performance was degraded as part of our evaluation. Figure 10.7 (top) illustrates the Garmin ForVü navigation Sonar in both the nominal and degraded states, while Figure 10.7 (middle) illustrates the nominal and degraded states for their LiveVü navigation Sonar. The EchoPilot forward-looking navigation Sonar imagery shown in Figure 10.7 (bottom) illustrates

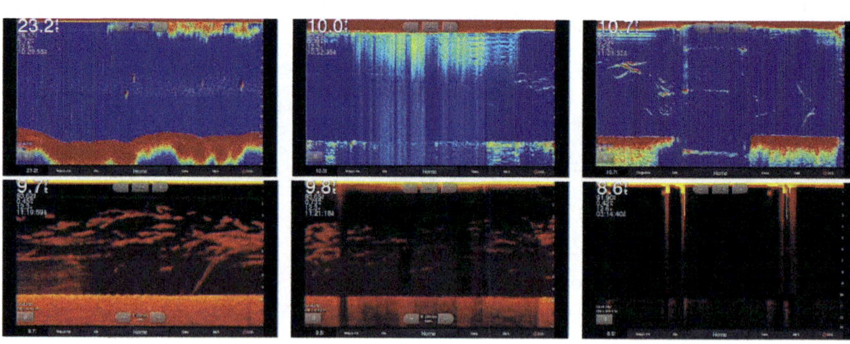

Figure 10.6 Nominal and Degraded Fish Finder (top) and Chirp Sonar (bottom) Images.

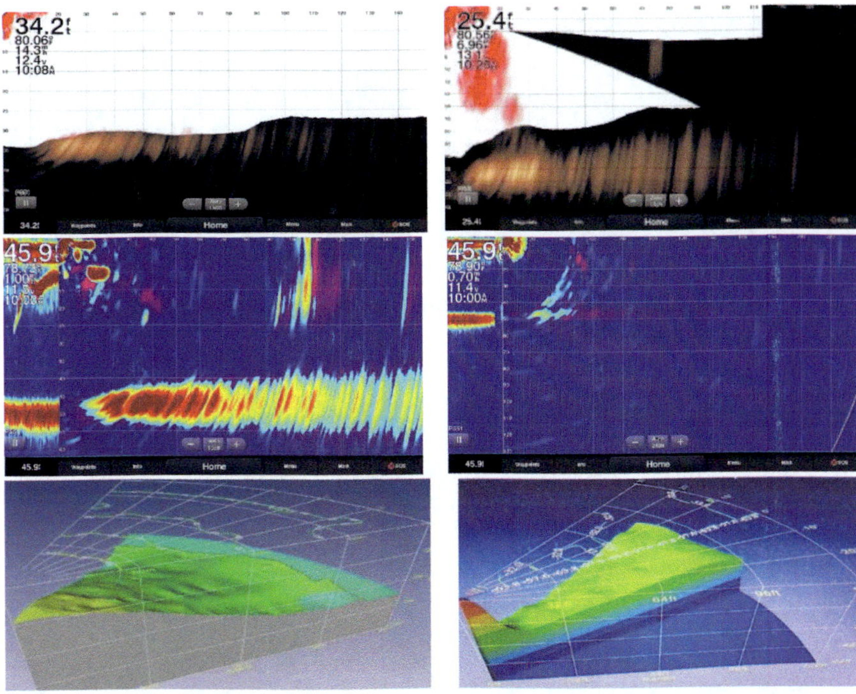

Figure 10.7 Nominal and Degraded ForVü, LiveVü and EchoPilot Navigation Sonar Sensor Images.

sensor degradation where the starboard of two sensors mounted athwartships on the bottom has become completely disabled due to bottom fouling. The port sensor was also fouled extensively but resulted only in slightly degraded performance.

Figure 10.8 provides an example of degradation using the FarSounder Argos 350 forward-looking navigation Sonar resulting from an obstruction in the field of view of the transducer. Additional cases of degradation were attempted using bubble screens and other techniques in the field of view. However, rather than resulting in degraded operation, the Sonar accurately depicted the phenomena with little observable degradation. Further examination of resilience to degradation effects on this sensor is warranted.

10.5.5 Lidar

The Oster OS2-128 long-range Lidar system produces an image spanning 360° with the vessel centered in the middle. The top image in Figure 10.9 illustrates a nominal operation where there is a natural blind spot on the navigation light array (outlined in the blue circle, left) directly below the field

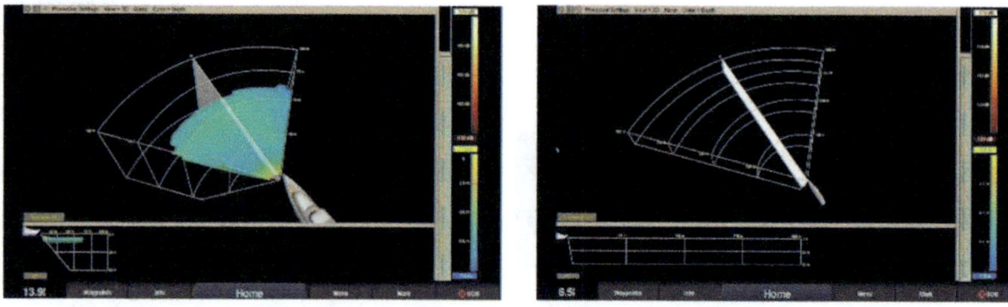

Figure 10.8 Nominal and Degraded FarSounder Navigation Sonar Sensor Images.

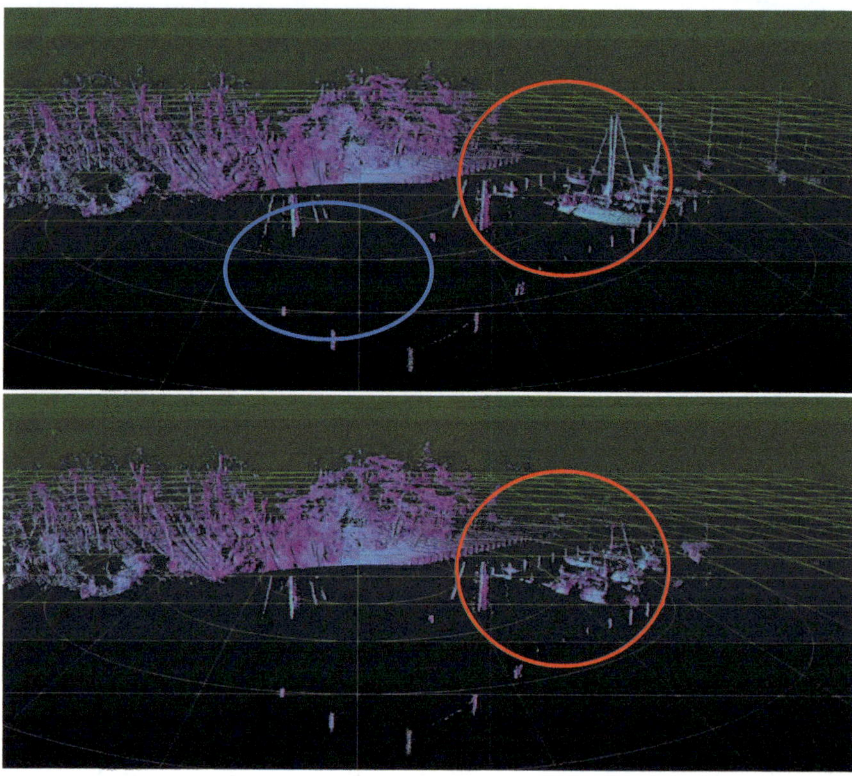

Figure 10.9 Nominal and Degraded Lidar Images.

of view of the sensor. The imagery in the red circle (right) illustrates the area that has been degraded in the bottom image as a result of an obstruction placed in the laser propagation path that causes a shadow effect where detail is lost.

This phenomenon was observed using several different types of obstructions of differing sizes, ranging from the smallest at 3 × 7 inches, which resulted in the degradation shown in Figure 10.9, to 24 × 36 inches, which caused massive shadows. All episodes of degradation within these obstruction sizes were routinely detected, while small sizes were marginally detectable depending on the background content.

The images shown in Figure 10.10 (top) represent the effects of interference caused by the reflection of laser energy from nearby structures back into the sensor from where it originated. The actual degradation manifested itself as a rapidly blinking display from 0.5 to 2 times per second where a large portion of the image was missing. This type of sensor degradation was consistently detected without any difficulty. Degradation also occurred, as shown in Figure 10.10 (bottom, left), resulting from the effects of heavy rain very similar to that exhibited by the FLIR sensor shown in Figure 10.3. Figure 10.10 (bottom, right) illustrates the effect of an insect (wasp) landing on the sensor resulting in horizontal lines within the red circle.

Additional degradation was attempted by inflicting strikes on the sensor with a 30 mW blue laser (450 nM), 25 mW green laser (532 nM) and 3 mW infrared lasers (780 nM and 905 nM). The green laser caused the Lidar system to freeze the image and had to be restarted. The blue laser seemed to cast shadows from the direction of origin, which were detected. The infrared lasers caused interference lines as well as occasional blinking of the display, which were also detected.

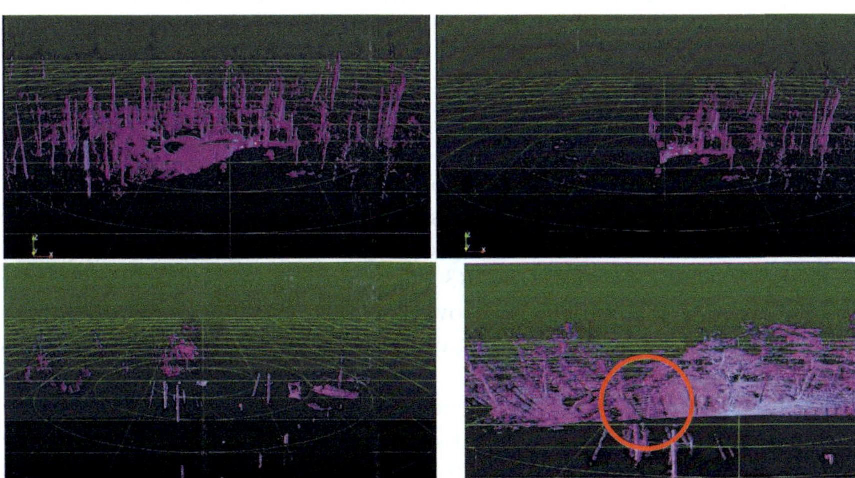

Figure 10.10 Lidar Degradation due to Laser Interference (Top), Heavy Rain (Bottom, Left) and Insect Landing on Sensor (Bottom, Right).

10.5.6 Global Positioning System (GPS)/GNSS

Various efforts were undertaken to achieve DoS attacks on the Garmin GPS/GNSS, without success. The problem stems from the ubiquitous nature of modern displays where this capability is built into each display, and each is connected to a common NMEA-2000 bus. Our research vessel testbed has four Garmin 8616/8612xsw displays, as shown in Figure 10.1 (top, center) that can function without an external antenna. Each time one or more units were disabled through physical obstruction and electrical isolation of satellite signals there was a seamless transition to other active devices. Attempts to simultaneously isolate all receivers proved to be impractical. Consideration was given to attempting to electrically interfere with the satellite signals using our onboard SDR transceiver generating low power (<100 mW) complex signals. However, our geographic proximity to several large metropolitan airports and veritable nonstop activity of nearby commercial, general aviation and military aircraft within the line of sight was determined to be too risky and this approach was not attempted. However, the effect of such transmissions on various different types of GNSS receivers in close proximity would certainly prove interesting.

An alternative approach used our legacy Furuno GP-37 GPS receiver that was not connected to the NMEA-2000 bus, but rather to a fully independent NMEA-0183 bus shared by the Advanced Navigation INS. In this case we were successful in isolating this receiver by insulating the antenna from satellite signals using metal screening. This resulted in the activation of the WAAS/DGPS ERROR! Alarm, the alarm icon appearing on the user interface screen and reporting the condition over the NMEA-0183 bus, which was detected. Degradation was also detected using bottom-depth contour tracking in conjunction with INS experiments.[9,10]

10.5.7 Inertial Navigation (INS)

The Advanced Navigation GNSS Compass/INS provides an accurate dual antenna GNSS-based heading that is not subject to magnetic interference and can maintain accurate heading during GPS outages. Positioning data is provided across the NMEA-0183 bus. During a loss of heading fix the GNSS Compass will continue to provide accurate heading for up to 10 minutes using its gyroscopes. A heading error will slowly accumulate during this period at a rate of approximately 15 degrees/hour. During a loss of position fix the GNSS Compass will continue to provide accurate position and velocity for up to 2 minutes using its inertial sensors.[11] This receiver was successfully isolated by insulating the antenna from satellite signals using metal screening. Real-time tracking of actual GNSS position was accomplished with the Garmin GPS using the NMEA-2000 bus, while tracking of INS positioning was accomplished over the NMEA-0183 bus.

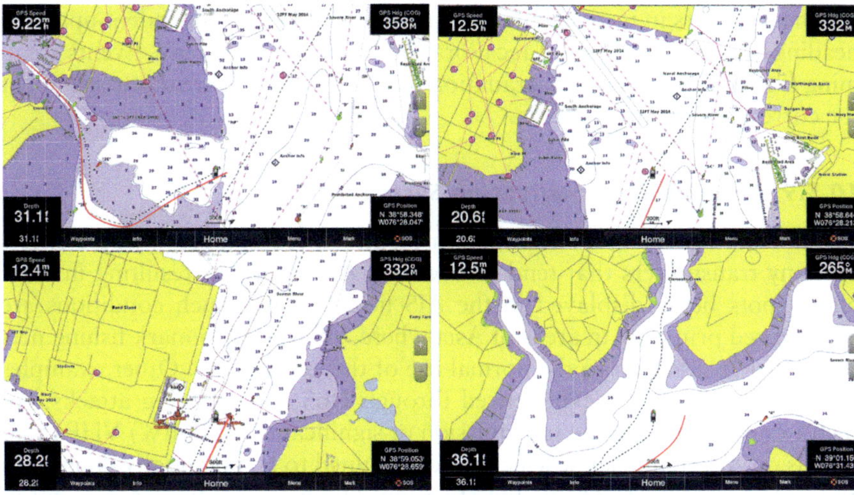

Figure 10.11 Inertial System Degradation over Time without GPS/GNSS Inputs.

Figure 10.11 illustrates the progression of INS error over time along an approximately 12-mile track, with INS positioning depicted by the red line while vessel positioning via GPS is shown by the black dashed line with the boat icon. The black dashed line track to the right of the vessel track is an artifact remaining from a previous voyage and should be ignored.

GPS/GNSS signals were disrupted upon voyage commencement. The top-left image depicts the beginning stage of the voyage at a point approximately 10 minutes from leaving the pier where the INS was still providing accurate positioning and heading information. A series of turns during this period appeared to hasten the onset of positioning error while heading information remained relatively accurate. The top-right image progressing into the bottom-left image depicts a steady positioning error that continued relatively unchanged while navigating a relatively straight line. At a point close to voyage termination after approximately 45 minutes, the track shown in the bottom right image shows the INS course has degraded significantly.

Detection of INS degradation was accomplished as a GNSS reception outage was reported by the system over the NMEA-0183 bus. Secondly, positioning error was detected through bottom depth and contour tracking that determined local depths at the vessel position did not match those reported by the INS but indeed corresponded to a position hundreds of feet away from the vessel.[8,9]

10.5.8 Heading Sensor

Degradation to the heading sensor was introduced through the application of electromagnetic fields in close proximity to the sensor. This resulted in significant errors in instruments such as Radar and ENC dependent upon the heading sensor and required recalibration of the ENC display. Events stemming from heading sensor degradation were detected and are discussed under the corresponding instruments.

10.5.9 Automatic Identification System (AIS)

The primary mode of AIS degradation we attempted to detect was the effects of many transmitters that tend to overload AIS receivers, resulting in large blank spots being displayed in the AIS field of view. Such conditions are experienced primarily in areas of Asia where AIS is used to mark fishing nets and buoys in addition to the normal use of this technology. Other examples of degradation can be experienced through DoS and spoofing attacks. All such examples require the transmission of significant (>100 mW) VHF radio signals on AIS frequencies, which we determined to be potentially dangerous and outside the scope of our investigation. However, we were successful in achieving a denial-of-service attack using low-power (<100 mW) VHF transmissions of complex waveforms that resulted in a blank AIS display, which was consistently detected.

10.5.10 Engineering Sensors

There are a wide variety of engineering sensors and data types that are interconnected via the NMEA-2000 bus network.[12] Similar information is available on the NMEA-0183 and other bus networks. However, their data communication protocols and formats can vary widely but the concepts involved are applicable to many of the prevailing instrument and sensor bus technologies found in commercial maritime, naval applications such as the U.S. Navy Unmanned Maritime Autonomy Architecture (UMAA),[13] as well as for aviation, automotive and railroad use.

The primary focus was centered primarily on data that was available from the various sensors utilized in our experiments, including:

- AC Power
- DC Power
- Depth
- Engine
- Environment
- Fuel Flow
- GPS
- Heading
- Navigation
- Pressure
- Rudder

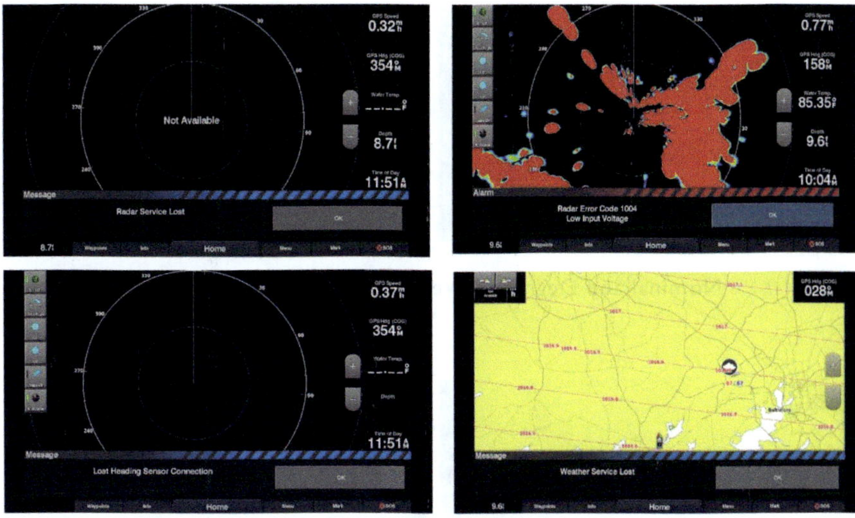

Figure 10.12 Degraded Engineering Sensor Detection via NMEA-2000 Data Bus.

- Speed
- Tanks
- Temperature
- Time
- Vessel
- Wind

Sensors from the above list were evaluated in the detection of sensor degradation. Several of these are shown in Figure 10.12. Moving from top left to bottom right, these images denote sensor degradation to various degrees reportable by the sensors themselves over the NMEA-2000 bus and include instances where Radar service was lost, low DC voltage levels were detected, the connection with the Heading Sensor was lost, and satellite weather service was no longer available. This means of discerning sensor degradation and anomalies over the NMEA-2000 bus complements the three-domain approach (Time, Frequency and Spatial) and can provide greater clarity and specificity that would enable not only the detection of sensor degradation but also the identification of what types of degradation exist and possibly the means to overcome the degradation.

10.5.11 Weather Sensors

Degradation of onboard weather sensors was performed by introducing external stimuli and situations that would interfere with normal operation.

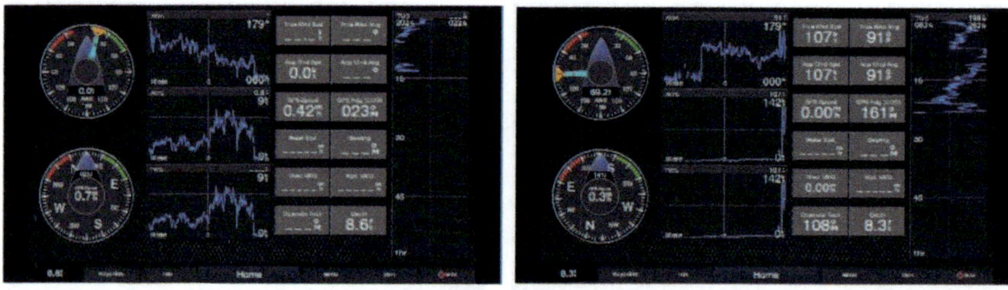

Figure 10.13 Nominal and Degraded Weather Sensor Images.

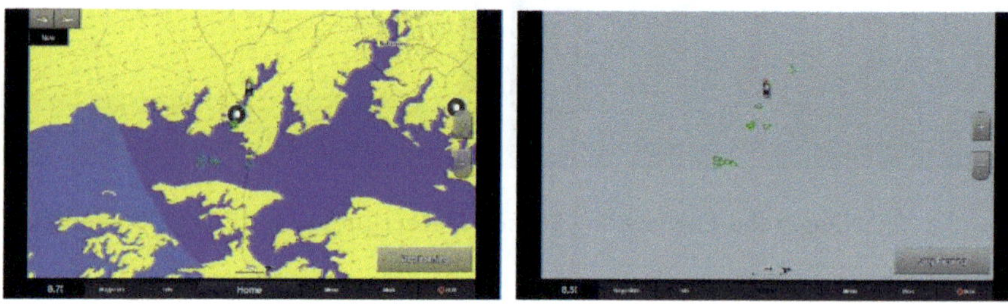

Figure 10.14 Nominal and Degraded Satellite Sensor Images.

This includes placing a box over the weather sensor, spraying the sensor with streams of water, applying a heat gun to artificially increase ambient temperature and subjecting the sensor to high winds from a local source that included a fan. Figure 10.13 illustrates the effects of a stream of water that resulted in maximum readings in wind speed and relative humidity, abnormal temperature readings and other measurements that were uncharacteristic and not representative of the physical environment. Anomalies were readily detected related to water, temperature and wind while isolating the sensor from the environment. However, wind anomalies were not readily discernable from a calm day without wind and with steady temperatures.

10.5.12 Satellite Sensors

Degradation of reception of satellite transmissions for weather and other services was performed by isolating the receiving antenna through the use of metal shielding. The loss of signal resulted in the degradation shown in Figure 10.14 which was easily detected.

Note that in the absence of signal the weather display became blank. However, the AIS overlay still continued without interruption.

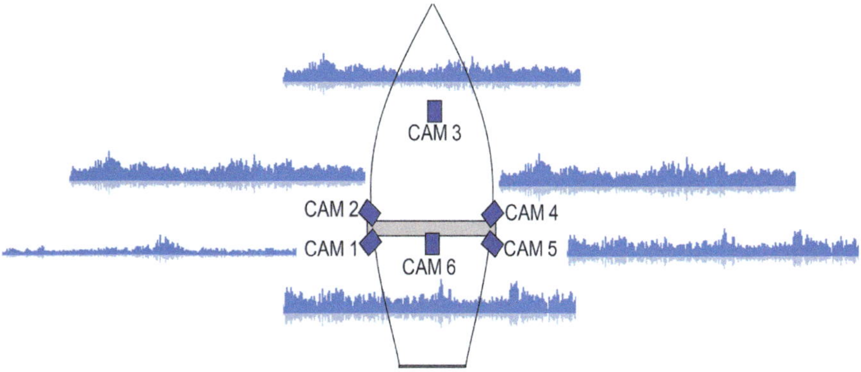

Figure 10.15 Microphone Placement and Audio Waveforms.

10.5.13 Microphone

Audio sensors are vital to ensure compliance with COLREGs Rule 5 requirements to maintain a proper look-out by (sight and) hearing.[14] Degradation may result in a lessening or complete inability to hear sounds such as horns, whistles, sirens and bells emanating from other vessels as well as aids to navigation that include buoys and lighthouses. This experiment utilized microphones built into the six onboard cameras oriented at 60-degree angles surrounding the vessel. The Camera 1 microphone illustrated in Figure 10.15 was degraded using a covering resulting in lower sensitivity as illustrated by the reduced and distorted sound recorded on the corresponding waveform. The degradation illustrated for Camera 1 and similar events directed at the other microphones were consistently detected with directionality determined.

10.5.14 Electronic Navigation Chart/Electronic Chart Display System (ENC/ECDIS)

Sensor degradation as may apply to ENC and ECDIS was investigated and determined that the effects would very likely appear as errors and inconsistencies in the chart display. However, such events are most likely to originate from the navigation, AIS and other sensors that feed data to the ENC. Therefore, we determined the best means to detect degradation would be directly from these sensors themselves rather than through the ENC which is secondary to the sensor. Opportunities to further investigate ENC/ECDIS degradation detection may prove to be useful.

10.6 Image Snippets for Remote Analysis

In each case where sensor degradation detection was achieved a small snippet of degraded sensor imagery is captured that could be sent to a Remote

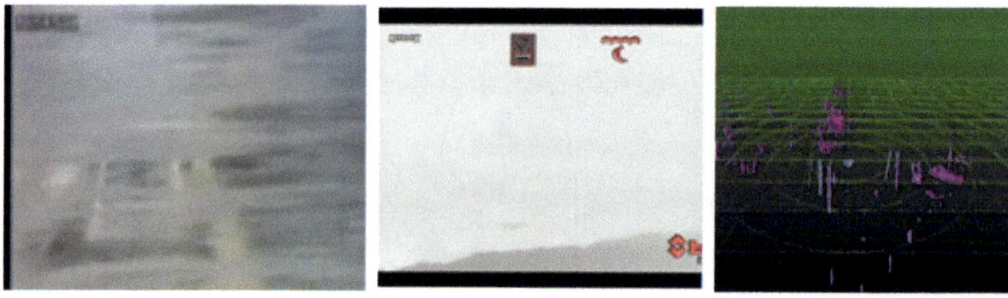

Figure 10.16 Image Snippets to Return to Remote Control Center for Analysis.

Figure 10.17 Transmission of Snippets via Email Using High Frequency (HF) Radio.

Control Center (RCC) via low bandwidth communication channel for human analysis at a RCC as well as machine analysis. Snippets, samples of which are shown in Figure 10.16, are 160 × 145 pixels in size with each requiring memory from 62k and 118k of memory, without compression. Compressed snippets would result in even smaller memory requirements.

This is an example of just one way to present this information in a manner useful to staff at a remote-control center. Figure 10.17 illustrates an example of how the communication of snippets was accomplished using high-frequency (HF) radio communications whereby imagery from three sensors was forwarded from the research vessel directly to our offices for use by a remote operator. HF radio was selected to demonstrate this capability over a low-bandwidth communications channel that can span long distances assuming that satellite communications were unavailable. Other methods to achieve the same result using broadband communications are readily available.

10.7 Remote User Interface

A user interface, illustrated in Figure 10.18, was created for the display of sensor status and sensor image snippets at an RCC.

Sensors that are currently not configured into the active vessel architecture are grayed out while active sensors are colored in red, orange, yellow and

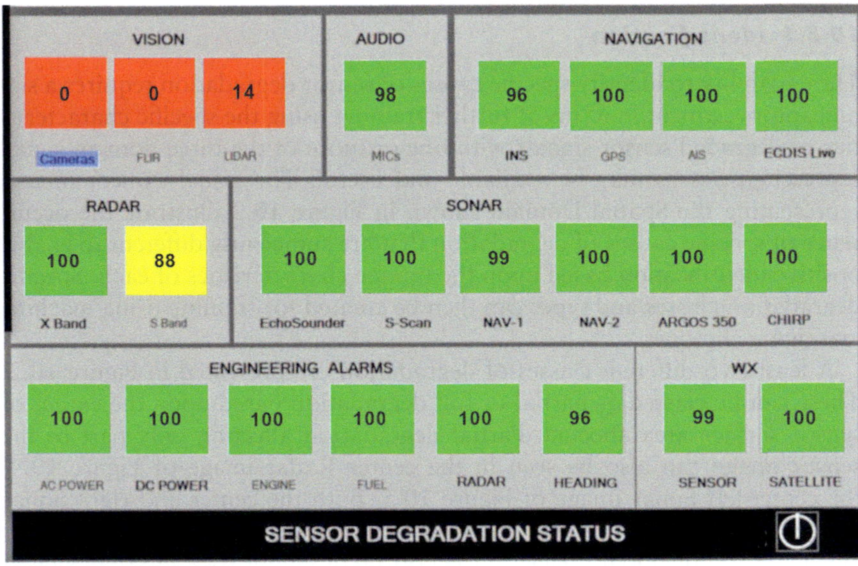

Figure 10.18 User Interface for Remote Display of Sensor Status.

green to illustrate general sensor degradation status. A number from 0 to 100 is also shown to indicate degree of sensor degradation.

10.8 Sensor Degradation

Detection of the existence of sensor degradation is based upon determination that the characteristics of the examined Frequency, Time and/or Spatial Domain sensor signal representations differ from the characteristics associated with nominal sensor operation as embodied within the training data sets used during decision-making software development. This process is tailored to the data characteristics of specific sensor types and models with respect to degrees of variation from nominal, and nothing else. All examples of degradation presented in this chapter were detected in this manner.

The training requirements for detection presume a very large volume of data across the three domains over long periods of time, preferably with continuous learning so that the training data will continue to improve throughout the useful life of the sensor(s). This data must contain representative samples of nominal operating conditions for all sensors involved. Examples include day, night and twilight; calm to rough seas, fair to stormy weather, empty screens to ones filled with targets and shallow to deep waters. This is an area where deep-learning AI takes a lead role to ensure that robust and comprehensive training is achieved with minimal false indications of degradation.

10.8.1 Identification

The capability to identify specific causes of sensor degradation requires a significant investment in terms of further training using the specific characteristics of degraded sensor signals with one or more of the three domain signal representations as may be available and useful. The visual camera images representing the Spatial Domain shown in Figure 10.2 illustrate the occurrence of specific causes of degradation that are sufficiently different to enable positive identification based upon the unique characteristics of each. Specific degradation classes and types can then be created for training using machine learning techniques.

A least two different classes of degradation are presented in Figure 10.2. These can be classed as partial or full degradation based upon the extent of sensor surface area affected. Partial degradation affecting only part of the sensor image can also be seen in the center Radar image of Figure 10.4, the center-left Sonar image of Figure 10.5, both the center and right Sonar images of Figure 10.6 and the Sonar images on the right side of Figures 10.7 and 10.8; and the Lidar images at the bottom of Figure 10.9 and right side of Figure 10.10. Full degradation that affects 50% or more of the sensor image can be seen in many of the other figures. Degradation shown in Figure 10.2 can be attributed to damage, obstruction, obscurant and natural-fog and natural-rain in the following order as shown in Table 10.2:

The examples cited represent identification of the causes of degradation based only upon the unique characteristics presented from Spatial Domain perspectives. This alone may be sufficient to accurately and consistently identify particular types and sources of degradation. In cases as shown in Table 10.2; frames 4, 7, 11 and 12 of Figure 10.2; additional insight may be needed to determine proper classification and type. In such cases unique characteristics associated with each degradation may also exist within the Frequency Domain as the properties of each causal agent may filter specific frequencies of light entering the camera and possibly introduce new frequencies that are not naturally present. These inter-domain correlations can help to provide greater specificity in identification. Examination of Time Domain characteristics can help to identify the exact timing degradation began, whether it occurred quickly or gradually, the extent of degradation, and

Table 10.2 Identification of Classes and Types of Degradation Shown in Figure 10.2

1. PARTIAL, DAMAGE	2. FULL, OBSCURANT	3. FULL, DEBRIS
4. FULL, OBSCURANT	5. FULL, FOG	6. PARTIAL, FOG
7. FULL, RAIN	8. FULL, DAMAGE	9. FULL, OBSCURANT
10. FULL, OBSCURANT	11. FULL, OBSCURANT	12. FULL, DAMAGE

whether it continues unchanged or if the sensor is recovering from the effects of the degradation agent through evaporation, dislocation or other scenario.

10.8.2 Remediation

The capability to overcome the effects of sensor degradation may be achieved through the use of sensor fusion techniques to derive the contents and meaning of degraded data provided by one sensor by examining data from other sensors to arrive at the same or similar conclusions. Such an approach can provide greater resiliency in decision making by using sensor data obtained from different portions of the electromagnetic spectrum (Frequency Domain) over the same periods of time (Time Domain) while providing fixed reference points for comparison between different sensor data products (Spatial domain). Further, such an approach may provide a second source for information that is normally gathered from another sensor. Several examples are provided within this paragraph to demonstrate the potential of this approach to insure greater resilience in sensor degradation detection performance.

10.8.2.1 Visual, Infrared, Lidar, and Microphone

The fusion of multiple sensor data streams can be used to reinforce the conclusions reached by individual sensors and to eliminate ambiguities as to which sensors may actually be degraded. Figure 10.16 illustrates snippets of visual and infrared cameras as well as Lidar that have been degraded as a result of heavy rain showers. Such data can be supplemented with audio obtained from microphones to further reinforce conclusions. The simultaneous reporting of degradation by sensors that utilize dissimilar methods at different points within the radio frequency spectrum can provide increased confidence in degradation detection performance.

10.8.2.2 Visual and Weather

Visual camera imagery can reduce ambiguity in determining if sensor measurements accurately depict the external environment. In the case of the weather sensor illustrated in Figure 10.19 showing winds measured at 107 knots, a simple comparison with live imagery indicates a sea state much more appropriate to calm rather than hurricane conditions. The introduction of such capabilities can be accomplished through training of visual camera neural networks to readily discern between different sea states at various Beaufort Wind Force Scale conditions.

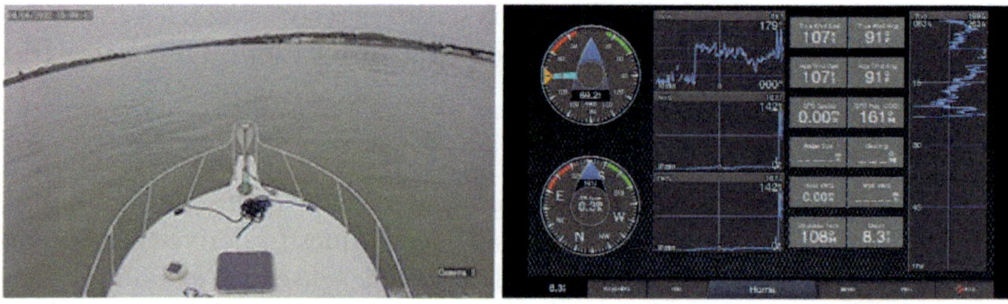

Figure 10.19 Example Use of Sensor Fusion to Remediate Sensor Degradation.

10.8.2.3 Echosounder and ENC/ECDIS

Several references have been made throughout regarding the use of bottom depth and contour tracking as a means to detect degraded Global Positioning System and Inertial Navigation System sensors that provide errors in geographical positioning.[9,10] ENC/ECDIS hydrographic data is georeferenced to position by latitude and longitude and can be used as an alternative source of information to the echosounder regarding depths at any particular location. Our approach utilizes these data to make comparisons between charted depths and live echosounder readings to help determine whether degradation may exist in either position sensors or the echosounder. In well-charted areas with recent hydrographic survey and adjusted for tide information these measurements should be relatively close. In areas lacking recent hydrographic survey or after significant storms that may cause shoaling and changes in seabed configuration these values may vary widely warranting notification to navigation processes and remote controllers that caution should be exercised.

10.8.2.4 AIS, Radar, and Visual/Infrared

Many vessels are not equipped with Automatic Identification System (AIS) technology while others intentionally go dark by turning off their AIS transponders making it difficult to obtain a positive identification of vessels while underway. Oftentimes AIS signals are spoofed such that a vessel may appear at a location that is miles to thousands of miles distant from their actual location. Further, denial of service attacks on AIS sensors may also make them inoperable. All of these instances provide examples where AIS capabilities are degraded through intentional or nefarious action even though the sensors themselves are perfectly operational. In such cases it is possible to utilize Radar, visual cameras and/or infrared cameras in an attempt to resolve ambiguity as to whether an AIS sensor data is degraded or not. They can also be used to determine the presence or absence of a vessel and possibly to identify such vessels.

10.8.2.5 Microphone, Visual, Infrared, Lidar, Weather and Satellite

Sound information obtained from microphones may reinforce indications obtained from other sensors, or they may weaken and challenge the conclusions reached through analysis of other sensor data. For example, sound can boost assertions from visual and weather sensors that it is indeed raining heavily, or snowing, or encountering hail. Likewise, the presence of unusual sound or the absence of sound may indicate nefarious activity. In all cases where sensor degradation is indicated, the use of microphones strategically placed throughout the vehicle can help resolve ambiguities as to the source and extent of degradation that may exist.

References

1 Jim Myer. NMERPAC to look at mariner shortage. *Waterways Journal*. 2 September 2022. National Merchant Marine Personnel Advisory Committee. U.S. Coast Guard. www.waterwaysjournal.net/2022/09/02/nmerpac-to-look-at-mariner-shortage/

2 R. Glenn Wright. In-Stride Detection of Sensor Degradation. GMATEK, Inc. Final Report. Revision: A, 28 April 2022. Contract: N6833520G2005. Naval Sea System Command, Washington, D.C.

3 Wright, R. Glenn. *Signals Intelligence Automated Assessment of Test Capabilities*. IEEE Automatic Testing Conference. Washington, DC. September 2018.

4 John Boyd. New Security Technology Detects Malicious Cyberattacks on Drones, Cars and Robots. 26 February 2019. https://spectrum.ieee.org/new-security-technology-detects-attacks-on-sensors-controlling-numerous-applications-including-drones-cars-and-robots

5 Michael Koziol. New Ethernet Cyberattack Crunches Critical Systems. 15 November 2022. https://spectrum.ieee.org/cyberattacks

6 A. Aboulian, et al. (2019) NILM dashboard: A power system monitor for electromechanical equipment diagnostics. *Trans. on Industrial Informatics*, 15(3), 1405. DOI: 10.1109/TII.2018.2843770

7 DARPA Assured Autonomy Project. www.darpa.mil/program/assured-autonomy

8 *Fouling Ratings (FR) in Order of Increasing Severity*. Naval Ship's Technical Manual. Waterborne Underwater Hull Cleaning of Navy Ships. S90860CQ-STM-010, rev. 5. Chapter 081. Naval Sea Systems Command. 1 October 2006. Table 081-1-1.14.

9 Wright R. Glenn and M. Baldauf (2016). Hydrographic survey in remote regions: Using vessels of opportunity equipped with 3-dimensional forward-looking sonar. *Journal of Marine Geodesy*, 39(6), 439–357. DOI 10.1080/01490419.2016.1245266

10 Wright, R. Glenn. *Enhanced MASS Situational Awareness Using Virtual Navigation Aids*, Second International Conference on Maritime Autonomous Surface Ships (ICMASS 2019). Trondheim, Norway, 13–14 November 2019.

11 Advanced Navigation. GNSS Compass Reference Manual. Version 1.0. 23 January 2018, p. 36.

12 Maretron. Supported Display Data Types DSM570 NMEA 2000® Multi-function Color Graphic Display, User's Manual. p. 260, Revision 1.8.3.

13 U.S. Navy. Unmanned Maritime Autonomy Architecture (UMAA). PMS 406. Unmanned Maritime Systems. Naval Sea Systems Command. www.navsea.navy.mil/Portals/103/Documents/Exhibits/SAS2021/SAS-2021-UnmannedMaritimeSystems.pdf

14 COLREGS, Rule 5. Look Out. Convention on the International Regulations for Preventing Collisions at Sea, 1972, (72 COLREGS). Navigation Rules COMDTINST M16672.2D. United States Coast Guard.

Chapter 11

Next-Generation Sensing

The primary difference between conventional ships and unmanned and autonomous ships is the latter incorporate enhanced sensor suites, sensor data fusion technology, computers and communication systems that allow them to operate without seafarers onboard. These technologies that benefit efforts towards the automation of ship operations continue to be integrated onto the bridges of conventionally staffed vessels to assist seafarers in the performance of their duties.

One outcome of these events is the orders of magnitude increase in the volume of imagery, signal and digital data generated by sensors on ships in recent years that drive enhancements to hardware and software leading to new and improved processes that make better use of these resources. The promises of innovation in the scope of onboard sensors, their uses and applications, and the data they generate have only just begun to be tapped. These provide new perspectives from which enhanced situational awareness can benefit all engineering functions and increase the safety of navigation, cargo handling and transportation; and passenger safety, comfort and services. Examples include revealing actual and pending deficiencies in sensors and ship systems, predicting and avoiding nearby vessel traffic with greater precision, improving ship routing and vessel efficiency, detecting hazards to navigation, expanding the roles of sensors to achieve new capabilities, and considering previously unknown mechanical, electrical and environmental factors associated with vessel operations.

This chapter attempts to highlight several advances in sensor technology as apply to present, developing and entirely new generations of conventionally manned, unmanned and autonomous vessels. These sensor technologies and new inventions must be considered in terms of functional, practical, legal and ethical impact on future enhancements to the International Maritime Organization (IMO) Safety of Life at Sea (SOLAS) Convention, the Polar Code, the International Convention on Standards of Training, Certification and Watchkeeping for Seafarers (STCW), Convention on the International Regulations for Preventing Collisions at Sea (COLREGs) and other international codes, conventions, guidance documents and practices. Development

DOI: 10.1201/9781003377900-11

of these enhancements and their implementation must be assisted by the many maritime nation-states, shipping companies and labor organizations, classification societies, insurers and non-governmental organizations.

11.1 Ship-Based Drone Vehicles

There has been a massive increase in the use of drone technology throughout all industries since 2016 upon the granting of broad exceptions by aviation regulatory authorities allowing their routine operation by companies for many uses. Total global shipments reportedly will reach 2.4 million in 2023 and are increasing at a 66.8% compound annual growth rate.[1] This same report cites Goldman Sachs forecasts of total drone market size being worth $100 billion, supported by the growing demand for drones from the commercial and government sectors. Drone use by the maritime industry is following this same trend with many applications related to logistics and resupply at sea as well as other uses for man overboard search and rescue, hull and cargo inspection, scouting for navigation in ice and hazards to navigation.

One study published in 2020 supported by a North Sea Region EU grant cited three areas where drone use would likely become accepted practice by 2025: multipurpose drones for vessel navigation support, document delivery and man-overboard rescue.[2] Although the timeline cited may be subject to debate, their conclusions are generally in line with current industry events.

11.1.1 Logistics and Resupply at Sea

Rather than simply focusing on the delivery of documents, there is great interest in drone use for the delivery of supplies for ships underway as well as at anchor. One example occurred in January 2016 when Maersk Tankers began drone testing by making a flight of 250 m from one of its barges to a tanker and drop off a package, and again in March 2016 upon completing drone delivery to a ship to initiate a wider program to test whether unmanned aerial vehicles (UAVs) can be used and implemented for supplying vessels with spare parts, mail or medicine.[3,4] In the latter case, a package was successfully dropped from a height of 5 m above deck level. The use of small, lightweight UAVs to deliver small packages was seen as a means to eliminate the need for more conventional means of delivery as a way to cut costs.

In another test in 2019, Wilhelmsen Ship Service and Airbus completed a trial shore-to-ship drone deployment to a vessel at anchor in Singapore.[5] This test originated from the Marina South Pier with a cargo of 3D printed consumables where the Airbus Skyways drone navigated autonomously along pre-determined aerial corridors in its 1.5 km flight to the Eastern Working Anchorage. The drone took off, landed on the deck of Pacific Offshore anchor handling tug supply vessel M/V *Pacific Centurion Swire*, deposited its 1.5-kg cargo, and returned to land in 10 minutes.

Similar efforts were undertaken by the U.S. Navy in 2021 using a small drone to carry lightweight cargo between vessels to eliminate the need for helicopter flights for a substantial fraction of its spare parts logistics.[6] Their plan is to use a commercial drone to carry up to 24 pounds of cargo for a distance of 65 miles, which would be sufficient to airlift about 80% of the Navy's critical parts cargoes.

These details are indicative of one aera in which interest in maritime drone use is flourishing. However, since this is a book on ship sensors, the rest of this paragraph will focus on drone UAV use as a means to extend sensor presence beyond the confines of the ship or vehicle to achieve other purposes.

11.1.2 Man Overboard Search and Rescue

The use of autonomous and remotely piloted air vehicles to locate people in the water and possibly to drop flotation devices and other equipment has long been a need amongst ship crewmembers and shipping companies alike. With advances in drone sensor technology and increased drone payload capacity, this need can now be filled and should take high priority for ship owners and regulators in amending the IMO Convention on Safety of Life at Sea (SOLAS) to mandate the installation and use of such equipment on board medium to large vessels.

Worldwide incident statistics indicate 386 cruise and ferry passengers and crew went overboard between 2020 and 2023.[7] In the United Kingdom, overall, 40% of man overboard incidents resulted in death as reported by the UK Marine Accident Investigation Branch (MAIB). They examined 308 man overboard incidents between 2015 and 2022 from commercial and recreational boats, both at sea and inland.[8] The highest fatality rate was within the fishing industry where, out of 58 reported man overboard incidents, 33 (56%) resulted in a fatality. The second highest fatality rate occurred within the recreational boating sector, where 144 man-overboard incidents occurred with 69 (48%) resulting in a fatality. Worldwide statistics are difficult to obtain, but one study indicates a general assumption that approximately 1,000 people fall overboard annually.[9] A good source in the literature dealing with this subject for the period between 2011 and 2018 was advanced by Gönel and Çiçek.[10]

There has been a great deal of research in developing and testing drone solutions to the man overboard problem. One example is where members of the United Kingdom Royal NavyX team tested a heavy-lift Minerva drone UAV in delivering potentially life-saving equipment if someone falls overboard.[11] Over a several-month period in 2021, the team tested remotely piloted systems which locate personnel in the water, drop life-saving equipment and hover over the location until rescuers arrive. This was accomplished using a Minerva T-150 at the Royal Navy's Diving School at Horsea Island, Portsmouth and also on a civilian boat using a Minerva T-80 drone. The system was able to be controlled to take off the moving vessel and fly out to sea, successfully locating a dummy in the water before returning to land itself on a mat attached to the boat's deck.

Remotely piloted and autonomous UAVs equipped with both visible light and infrared cameras can play an essential role in speeding up the detection and rescue of people who have fallen overboard. High-resolution imagery can provide details of partial human forms in the water that are recognizable by remote UAV pilots as well as artificial intelligence (AI)-based smart cameras during daylight and low-light conditions in a variety of sea states. Infrared cameras can supplement visible light cameras to also detect human forms in the water using their heat signature that provides a sharp contrast to the generally colder water. One researcher has also developed and publicly released specialized annotated datasets for training and testing AI-based detectors for this task. This MOBDrone benchmark is a collection of more than 125K drone-view images in a marine environment under several conditions, including different altitudes, camera shooting angles and illumination.[12]

11.1.3 UAV Use for Sea Ice Navigation

Drone UAVs have become very handy in extending the vision of crewmembers from the bridge to provide overhead views of sea ice forward of the vessel to plan efficient routes through pack ice. Operations have been conducted at opposite ends of the planet in both the Arctic and Antarctic. Two of many examples are discussed in which the utility of UAVs to provide assistance in ice navigation has been proven under operational conditions.

In 2015, the Antarctic Division of the Australian Department of Climate Change, Energy, the Environment and Water used a quadcopter drone to help navigate the research and supply ship *Aurora Australis* through the sea ice on its annual resupply voyage to Casey station.[13] Flown off the aft deck of the ship, the UAV provided real-time imagery of the sea ice conditions ahead of the ship to help the crew optimize navigation decisions. One interesting aspect of this experiment was the fact that the drone's compass was adversely influenced by the ship's thousands of tons of steel potentially causing the UAV to behave unpredictably. The problem was solved by disabling the drone's GPS and flying it in full manual mode enabling the drone to operate successfully and deliver the required video imagery. During the nine-day voyage the UAV made five flights voyage to Casey requiring eight minutes in the air for each flight to collect the necessary data. A fixed-wing drone was also used to survey the area around Casey station from the air, mapping 30 hectares (74.1 acres) in 20 minutes. Other observations included the UAV can be launched in minutes, costs little to operate and saves valuable cargo space as it needs only a 4 m × 4 m space for launch and recovery.

On a project managed by Fairweather, LLC of Anchorage, Alaska, in the summer of 2016 a month-long voyage was successful in retrieving large mooring (anchor) systems used for offshore drilling platforms from along the Beaufort and Chukchi Sea floors.[14] During this voyage, offshore tug/supply ship *Aiviq* used an Aerovel Flexrotor drone launched off the helideck to survey the vessel's path guiding it through the least amount of ice. The UAV flew five flights hovering at several thousand feet totaling 19 hours and guided the fleet through otherwise slow seas choked with ice, saving the ship more than two weeks in their schedule.[15,16] Also used on this voyage were Sonar and underwater Remotely Operated Vehicles (ROV's) to precisely locate the anchors' positions prior to the arrival of the anchor handling vessels.

11.1.4 Ship Inspection via UAV

In 2020, cargo claims consultancy WK Webster performed an initial damage survey of the container ship *One Apus* using a remotely piloted drone while en route as the ship transited Osaka Bay.[17] Company inspectors in Kobe, Japan, also conducted a remote drone survey upon arrival. The imagery was taken by two drones using ultra-high-definition cameras that clearly showed the scale and detail of the devastation on board.

The trend towards more autonomous operations is a continuation of an approach already being used in shipyards. In September 2015, the Remontowa Shiprepair Yard in Gdańsk Poland remotely inspected internal spaces of a ship following an overhaul.[18] By using a high-definition camera Remontowa was able to record video and take still images of one of the cargo tanks to examine the condition of protective coatings and detect corrosion and cracks. In June 2016, the UK Royal Navy used drones to identify defects in HMS *Diamond*'s exteriors and reach inaccessible areas. Such capability was seen as being able to reduce time and cost during maintenance periods and allow ships to diagnose their own defects while still at sea. In the United States, a quadrotor measuring 23 in wide and 12 in high gathered information relating to onboard fire-fighting on a former US Navy ship. The UAV was able to navigate a smoke-filled location to map areas and locate fire, providing a clear safety benefit. These three examples illustrate the untapped potential for smaller drones not only on large unmanned vessels but across all vessels of all types and performing various missions and functions. Another land-based mission is the use of drones to monitor vessel emissions into the air and water. This is being tried in Istanbul by the Metropolitan Municipality's Marine Services Directorate to monitor the Bosphorus by taking photographs and video footage. Under a proposal by the European Maritime Safety Agency (EMSA) and the European Space Agency, drones could also be equipped with specialized sensors to detect and classify specific types of stack emissions such as sulfur and carbon dioxide in the English Channel, the North Sea, the Baltic Sea and the Gulf of Bothnia.

11.2 Forward-Looking Navigation Sonar (FLS)

Although Forward-Looking Navigation Sonar (FLS) has existed for decades, this valuable and highly overlooked resource has been virtually ignored as a proven means to help prevent groundings, collisions, allisions and the resulting loss of life, property and large-scale destruction of the environment. This paragraph describes FLS technology and presents a case study of one of many similar accidents where it would have positively detected the hazard that existed, and bridge watchstanders could have been notified well in advance in sufficient time to avoid disaster.

An important aspect of basic and enhanced training in ship handling and navigation courses is the familiarization with and handling of equipment, including transferring knowledge and experience on the capabilities, functionalities and constraints of their use. Bridge navigational equipment encompasses many and diverse systems to support safe and efficient navigation and protection of the marine environment. The present situation is characterized by an increasing level of integration of sensors, technical systems, displays and sophisticated decision support systems combined with complex alerts to ensure sufficient situational awareness of the bridge team. However, despite the presence of such sophisticated systems, accidents in the form of groundings and collisions still happen. Regardless of the behavior, actions taken or not taken by the bridge team, the grounding of *Costa Concordia* with the bridge equipped with very modern Integrated Navigational Systems (INS) can be seen as another prominent case where alerts implemented in the most modern and highly sophisticated navigation systems failed to raise the attention of those tasked to take action and avoid an accident.[1]

11.2.1 The Problem of Blind Navigation

The COLREGs and STCW require an adequate watch be maintained.[19,20] However, there is presently no established means to maintain a watch below the waterline to directly detect the presence of hazards to navigation. A key assertion of this author's research is that bridge watchstanders are generally provided only with indirect and/or secondary information from which safety-critical decisions are routinely formulated. Although the tragic grounding of *Costa Concordia* provides a convenient and fresh example to illustrate this assertion, the problem has and continues to exist on a daily basis during virtually every vessel transit. For example, nautical charts and even most modern Electronic Chart Display Information System (ECDIS) provide navigation information as a secondary reference created by survey at some time in the past that was accurate at the time it was made but not necessarily reflecting actual conditions at the present time of passage. Radar indicates only those nearby targets from which the proximity to underwater hazards to navigation should be deduced. A traditional echo sounder provides depth directly below the keel, but no indication of depth directly forward of the bow. These and other methods such as GNSS merely infer positional relationships to geographic locations of interest and concern, but none are capable of directly detecting the physical underwater Hazard to Navigation (HtoN) consisting of Scole Rocks off Isle del Giglio, or any other such hazard prior to grounding. In addition, there are many floating hazards that cannot be charted such as drifting shipping containers, debris fields, whales, etc., having little or no presence on Radar, other types of Sonar systems, or other navigation sensors.

Another assertion is that the technology needed to detect hazards to navigation in real time at and below the waterline currently exists in the form of FLS and is available as commercial off-the-shelf equipment ready to be installed on vessels at the next scheduled drydocking. This does not imply, however, that this technology is sufficiently tested, capable of being integrated directly into existing navigation systems in its present form, or adequate training exists for its use. Merely adding yet another sensor and display system to an already complex bridge environment without adequate engineering, planning and training is likely to make matters worse rather than improving them, resulting in increased risks to navigation. This was illustrated by the introduction of Radar that led to the 1956 Radar-assisted collision between M/Vs *Stockholm* and *Andrea Doria*. Lack of proper training in the operation, use and interpretation of Radar equipment combined with a lack of procedures for implementing corrective actions based upon Radar indications created a scenario where established procedures for navigating in restricted visibility were not followed, resulting in the collision.[21]

The capabilities and limitations of FLS technology are discussed in terms of their scope and capability to detect hazards to navigation, along with its suitability in terms of vessel type, speed and installation requirements. Also covered is the complexity of using FLS for vessel navigation describing present capabilities as well as future requirements in terms of standards for INS and bridge alerts. This includes the introduction of an electronic Virtual Aids to Navigation (VAtoN) as a means to display hazards to navigation on electronic chart display and information systems (ECDIS). Integral to this process is the use of simulation technologies to test, verify and validate system processes, procedures and training requirements well in advance of the introduction of this technology to vessels by the introduction of carriage requirements through the IMO.

11.2.2 Forward-Looking Sonar (FLS) Technology

Active Sonar, commonly referred to as an echo sounder by the IMO, is used by vessels to determine the depth between the keel and the sea bed.[22] A variation of the echo sounder is FLS used to detect bottom features and objects within the water column forward of the bow. Despite its usefulness and the availability of this technology in the commercial marketplace it is rarely included within the ships' complement of navigation sensors.

The methods used for detecting bottom features, objects and soundings by determining range, azimuth and elevation information can generally be described as variations on transmitting a steerable Sonar signal ahead along the path of the vessel or transmitting a single ping from which snapshots of the environment are obtained.[23,24,25,26] A mosaic of the bottom topography and specific targets is then created as the vessel progresses on its course.

The range of FLS can extend from eight to twenty times the depth ahead, depending on bottom and target conditions. It is most effective when the bottom topography slopes upwards and when targets are large and consist of hard rock and/or coral that provide good acoustic signatures. FLS products are available that provide both two-dimensional (2D) as well as three-dimensional (3D) representations that provide a more realistic portrayal of the course ahead.

FLS systems have been developed with different capabilities supporting both autonomous underwater vehicle and vessel applications. Of those designed for use on vessels, most are intended for pleasure and small fishing boats. There are systems with range and resolution that make them suitable for use on larger vessels such as workboats, offshore service vessels, merchant and passenger vessels. However, operational constraints may create limitations on their usefulness. For example, effective range may be limited by tradeoffs in transducer design to minimize water resistance and drag. A summary and examples of several FLS systems, presently available on the market and each utilizing a different type of transducer, and their specifications are given in Table 11.1.

11.2.3 Approach for Estimating Advance FLS Warnings

As a very first step for establishing and further development of definitive parameters for the issuance of warnings to watchstanders using FLS, certain guiding principles may be cited to achieve an estimate useful for broad guidance on the subject. For example, a calculation of the relative time afforded a vessel to begin corrective action such as altering course, reducing speed, or otherwise reacting to an alarm generated by an FLS would be based, at minimum, upon the following factors:

Table 11.1 Comparison between Forward-Looking Sonar (FLS) Systems

Specifications	*FarSounder Argos 1000*[a]	*Furuno CH-270*[b]	*EchoPilot*[c]
Maximum Detection Depth	85 m (278 ft)	100 m (328 ft)	100 m (328 ft)
Maximum Detection Range (MDR)	1,000 m (3,200 ft)	800 m (2,500 ft)	200 m (656 ft)
Operating Frequency	61 kHz	180 kHz	200 kHz
Vertical Coverage	60°	180°	90°
Horizontal Field of View	60°, 90°, or 120°	360°	60°
Maximum Transmit Power	<1,500 W_{rms}	800 W_{rms}	Not specified
Angular Accuracy	1.6°	Not specified	Not specified
Roll/Pitch Compensation	±20°	Not specified	Not specified
Roll/Pitch Accuracy	0.5°	Not specified	Not specified
Maximum Vessel Speed	25 kts.	10 kts.	Not specified
Screen Refresh Rate	0.3–2 s	8 s	1–2 s
Advance Warning at MDR @ 10 kts.[a]	183 s (3.1 min)	139 s (2.3 min)	28 s (0.5 min)

Sources: EchoPilot 3D Forward Looking Sonar; www.echopilot.com/user/image/fls3d-brochure.pdf; Furuno CH-270 Searchlight Sonar; www.furuno.com/en/business_product/pdf/marine/ch270.pdf; FarSounder 1000 Navigation Sonar; www.farsounder.com/files/f31566_3d-sonar-brochure_3.0.pdf

Notes:

Calculations by author.

[a] United States Patent 8,717,847; Blake, 6 May 2014.

[b] IMO Resolution MSC.232(82): 2006, Adoption of the Revised Performance Standards for Electronic Chart Display and Information Systems (ECDIS).

[c] IMO Resolution A.1021(26), Code on Alerts and Indicators, 2009.

$$Advance\,Warning(\text{sec}) = \frac{MDR(m)\text{-}SRR(\text{sec})\text{-}APT(\text{sec})\text{-}WRT(\text{sec})}{Vs\left(\frac{m}{\text{sec}}\right)}$$

where logical arguments may be made to establish values for each of these parameters:

Speed of Vessel (Vs) = 10 knots
Screen Refresh Rate (SRR) = 2 sec, or as noted
Maximum Detection Range (MDR) = obtained from Table 11.1
Alarm Processing Time (APT) = 4 sec
Watchstander Response Time (WRT) = 5 sec
1 m/s = 1.9438 knots

Although some systems can perform at much higher speeds, a value of 10 knots was selected for the Speed of Vessel (Vs) to provide a common basis for evaluating the reaction time to FLS system alarms. A value of two seconds for Screen Refresh Rate (SRR) on the display was assumed based on system performance specifications. However, SRR for the Furuno CH-270 system may be as high as eight seconds or more as this system provides general coverage that extends beyond the area directly ahead of the bow. The value for Maximum Detection Range (MDR) is obtained from the performance specifications of the individual FLS units.

Alarm Processing Time (APT) is the speed at which FLS data can be evaluated by signal processing and alarm generation algorithms to determine whether a condition exists that breaches predetermined vessel-specific safety contours and depths. This includes draft, course, maneuvering capabilities, lateral clearance margins and other factors pertinent to safety of navigation. An assumption is made that such criteria for FLS are likely to be similar to those established by IMO for ECDIS.[27] Integral to this factor are time delays to prevent normal operating conditions from causing false alerts because of normal transients that may exist in the FLS data.[28] This can add an additional one to several seconds to ensure target persistence, eliminating the generation of an alarm due to a single occurrence or short duration or transient target. A value of four seconds was selected based on FLS refresh rates as well as estimates of processing times for interface and communications systems handling software that may be required by an Integrated Navigation System.

Watchstander Response Time (WRT) is that needed for the Officer of the Watch (OOW) to acknowledge an alert and take appropriate corrective action based on the nature of the alarm. This must take into account time lags necessary to assess rates of change in processes, such as changing vessel course against targets' movements.[29] The time required for the OOW to confer with other bridge watchstanders and lookouts and to issue orders to the helmsman must also be factored into this calculation. Add to this the effects of fatigue and various other human factor elements, one can easily see that this factor is the most subjective and imprecise in the equation. A value of five seconds was chosen in part based on the author's direct observations of bridge practices used on several vessels under similar conditions using ECDIS and Radar indications.

Using these criteria, the EchoPilot 3D FLS provided approximately ½-minute warning, the Furuno CH-270 provided approximately 2¼-minute warning and the FarSounder 1000 provided approximately a 3-minute warning when considering their maximum detection range. The speed of 10 knots may appear somewhat slow for most vessels while underway. However, if a vessel is operating in unknown, poorly charted, or known-hazardous waters it is prudent to increase safety margins by proceeding at a slower-than-normal pace.

It should be noted that such advance warning calculations generally provide "best case" scenarios under ideal conditions and that actual conditions and response times must be expected to reduce these margins – significantly in some cases. Actual conditions must also take into consideration both human and technological factors that can result in major deviations from these response times. Technological factors can include water turbidity, poor acoustic reflection qualities of potential HtoN and even growth on the hull that may reduce FLS sensitivity. Human factors can range widely from distractions on the bridge, unfamiliarity with the equipment and general lack of training, proficiency or currency in watchkeeping procedures.

11.2.4 FLS Installations

To achieve the most reliable and accurate performance FLS transducer(s) must be located on an area of the hull that is free of turbulence from obstructions located forward of the mounting position. Each transducer must be provided an unobstructed view both horizontally towards the bow and vertically from the waterline to the bottom to achieve the best accuracy and effectiveness.

The EchoPilot 3D Forward Looking Sonar uses two transducers mounted athwartships, one on either side of the bottom equidistant to the keel – ideally on the rear third of the vessel. The interior installation of these transducers, as visualized from within the hull, is illustrated in Figure 11.1 (left).

The FarSounder 1000 Navigation Sonar uses one transducer mounted on the bow with a fairing tube inserted within an existing or standard bulb, or as a separate installation. The exterior installation of this transducer as seen from ahead of the bow, is shown in Figure 11.1 (center and left).

The Furuno Searchlight Sonar, Model CH-270 (not pictured) uses a single transducer that can be lowered from and retracted into the hull at speeds above 10 knots. This unit may be mounted anywhere on the vessel where an unobstructed view is available horizontally through 360°.

Figure 11.1 FLS Transducer Installations.

Note: Dual EchoPilot transducers mounted athwartships (left) and single FarSounder bow-mounted transducers (center and right).

11.2.5 Integrated Bridge Alerts

The information available from the FLS can be useful in alerting watchstanders as to potential HtoN present in the path directly ahead of the bow. Visual indications seen on the FLS display are one form of alerting mechanism. However, this data may also be shared as part of an integrated approach using ECDIS as well as the Alert management, module C of an INS according to latest IMO Performance Standards. This, moreover, coincides with existing STCW training requirements and could utilize existing alarm mechanisms with which watchstanders are already familiar.

The capability to provide those warnings and alarms can be even independent of operator interpretation and be based upon analysis of the information obtained from the FLS. This appears to be available only from the FarSounder system. Taking into account the characteristics of the sea area (e.g., approaching coastal waters, navigating in ice, etc.), the operator can configure parameters within the area being observed by setting alarm values by depth, minimum range, maximum range and field-of-view angle width as illustrated in Figure 11.2.[30] The alarm volume is shown in red. An additional setting is the number of hits detected before triggering the alarm.

Figure 11.3 shows an example of FLS information integrated as an ECIDS overlay. For this prototype Human Machine Interface (HMI) of integrated FLS information, the three major requirements valid for display of Radar information in ECDIS are applied with respect to the new technology.

Taking the potentials of FLS technology and results from first experimental implementations it is concluded that watchstanders are provided both visual and audible notifications as announced via the centralized Central Alert Management (CAM)-HMI and supported by presentation in FLS target overlay in an ECDIS as suggested in the Figure 11.3.

Furthermore, analysis of information available from FLS can be used to trigger alerts with different levels of priority. In accordance with the definitions provided by IMO performance standards the lowest level of such

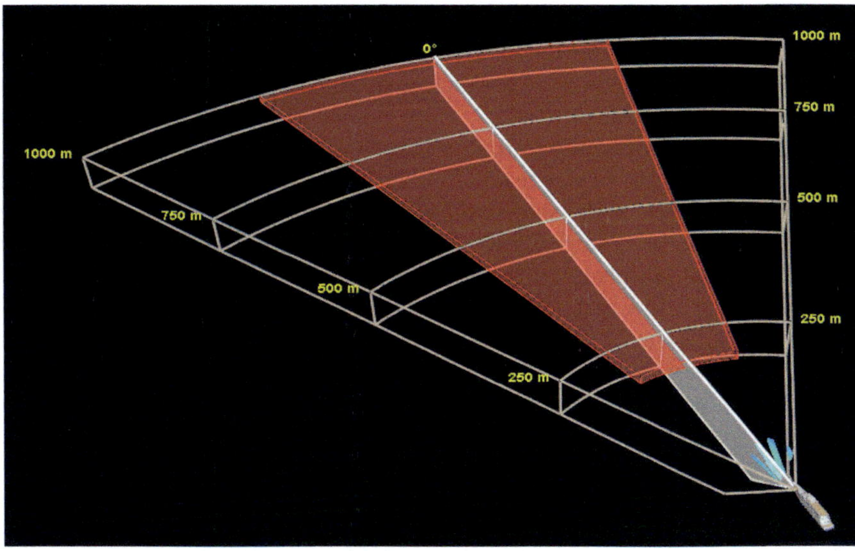

Figure 11.2 Forward-Looking Sonar Alarm Volume Setting. (image: FarSounder, Inc.).

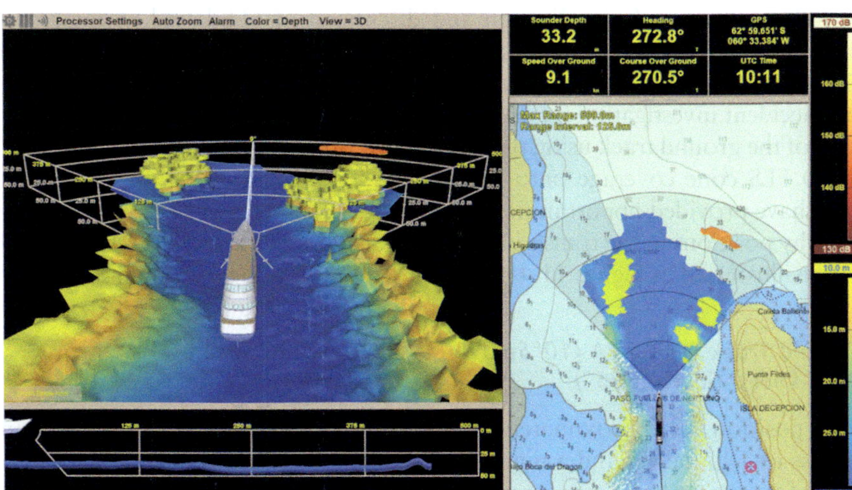

Figure 11.3 FLS Electronic Navigation Chart (ENC) Overlay. (Image: FarSounder, Inc.).

an alert is a caution to just raise awareness of the bridge team to a certain unusual situation. In respect to the integrated use of FLS information this could be a situation, where an obstacle is detected at a larger distance ahead, e.g., due to a cross-track error.

Triggering a warning, the second priority level of alert and requiring immediate attention by the bridge team could be linked to criteria within the usual maneuvering range of the ship to avoid contact to any detected HtoN. The time frame is to be configured as to allow the OOW to additionally check the FLS display but also match available ECDIS information (e.g., to proof approaching shallow water, a wreck or rock, island).

Finally, an alarm could be triggered characterizing conditions requiring immediate action when, for example, the ship is approaching the lower maneuvering limit (e.g., an emergency, hard rudder action would be required) to avoid contact with an FLS-detected object/HtoN. This last stage of an alert could be connected to the actual ship status and the ship's maneuvering capabilities. That information can be provided by dynamic predictions of the ship's future path for an emergency evasive maneuvers and check of the predicted track, for instance, against the available water depth.

11.2.6 FLS Information and Warnings

For the purposes of visualization and discussion of FLS-based warnings , a simulation of the circumstances encountered during the grounding of *Costa Concordia* in 2012 is used as a hypothetical case. Data are taken from the official accident investigation report[31] and simulation is attempted through extraction of the ground track using the AIS data record, then overlaying FarSounder 1000 FLS cone coverage onto the ground track to identify key events that may have provided opportunities to enhance situational awareness. Four noteworthy positions in the transit just prior to grounding identified as A, B, C and D are illustrated in Figure 11.4 and recorded in the below chronology.

Ground Track Position A (10°56'32" E, 42°20'41" N; Speed: 15.3 knots; Course 325°)

This position lies approximately 1,600 meters from Scole Rocks at a depth in excess of 100 meters and a mud bottom.[32] This location is outside the range of FLS to detect bottom information and is not likely to have provided any information of significance to navigation.

Segment A to B (Elapsed Time: 00:01:08, Rate of Yaw: 11°/min)

During most of this segment, the bottom depth would have remained in excess of 100 meters with a mud bottom. However, just prior to arriving at Position B, the maximum range of the FLS would have crossed the 100-meter

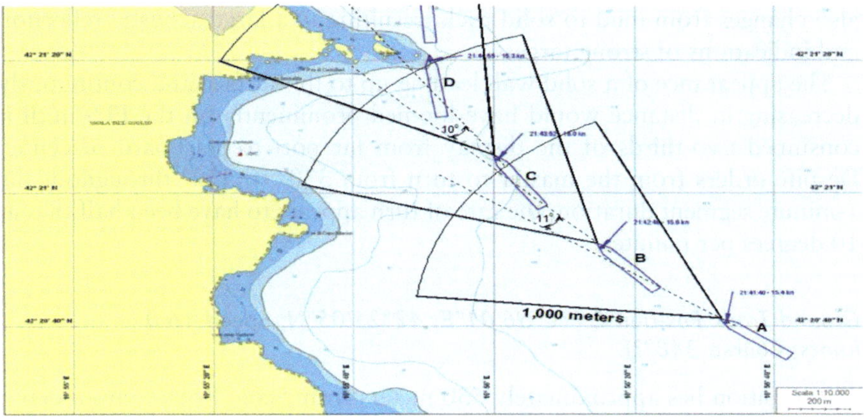

Figure 11.4 Ground Track of *Costa Concordia*, Annotated with Coverage for FLS Navigation.

Source: Marine Casualties Investigative Body, Cruise Ship Costa Concordia, Report on the safety technical investigation, p. 61, Ministry of Infrastructures and Transports (Italy).

bottom depth contour directly ahead and to the port of the centerline. It is possible that early indications of an upslope bottom may have been able to be detected. Note that during this segment, the rate of turn is approximately 11 degrees per minute.

Ground Track Position B (10°56'15"E, 42°20'52"N; Speed 15.6 knots; Course 335°)

This position lies approximately 1,100 meters from Scole Rocks at a depth in excess of 100 meters with a mud bottom. The 100-meter bottom contour is approximately 500 meters further along the course, with upslope bottom rising to the 10-meter bottom contour at the maximum range of the FLS off the starboard bow. The bottom is also transitioning from mud to rock, with a resultant increase in acoustic reflectivity of the bottom material.

Clear indications of the approaches to Scole Rocks would have appeared on the starboard bow on the FLS display. This probably would have initially alerted the Master to the existence of a problem since the appearance of Scole Rocks would have been expected off the port bow based on his passage planning.[33] Furthermore, the depth contour displayed on the FLS would clearly indicate insufficient depth in an unexpected location.

Segment B to C (Elapsed Time: 00:01:04, Rate of Yaw: 10°/min)

During this segment bottom depth is steadily decreasing across the range of the FLS from in excess of 100 meters to the surface. The bottom consistency also changes from mud to solid rock, resulting in a large acoustic reflection and indications of strong targets.

The appearance of a solid wall leading up to the surface and continuously decreasing in distance would have loomed prominently on the FLS until it consumed two-thirds of the display from far port to starboard of center. Despite orders from the master to turn from 330° to 350° throughout the 1-minute segment duration, the rate of turn appears to have been half that at 10 degrees per minute.

Ground Track Position C (10°56'01"E, 42°21'05"N; Speed 16.0 knots; Course 340°?)

This position lies approximately 530 meters from Scole Rocks with a depth of 95 meters just past the 100-meter contour with a rock bottom. The bottom contour would reflect an upslope bottom rising to the surface around 500 meters ahead. Less than one-third of the display off the starboard bow would now show clear water. There would be very little room to maneuver at this point, but it was still possible to avoid disaster.

Segment C to D (Elapsed Time: 00:01:03)

This is the terminal segment of the voyage. Bottom consistency is solid rock resulting in a large acoustic reflection and indications of strong targets. The appearance of a solid wall leading up to the surface and continuously decreasing in distance would have continued up to the point of impact.

Ground Track Position D (10°55'40"E, 42°21'21"N; Speed 15.3 knots; Course???)

This position lies approximately 30 meters from Scole Rocks with twelve seconds remaining before contact.

Hypothetic Course of Events – Applying FLS

Using a speed of 16 knots obtained from the above chronology as the value for Speed of Vessel (Vs) in equation (1), FLS could have provided approximately 1.8 minutes advance warning to the Master of *Costa Concordia* of the pending HtoN consisting of the approaches to Scole Rocks. The warning would have been issued when the vessel was just past position B, as shown in Figure 11.4.

Had FLS been installed and operational on the bridge as discrete navigational equipment without the benefit of inclusion within the INS, visual indications would have shown an unmistakable wall directly in the path of the vessel. Existing alarms available both integral and external to the equipment would have also been activated, further reinforcing the severity of the situation. Consistently, at all noted positions on the final approach (Figure 11.4: positions A through D), the FLS would have indicated clear water was present off the starboard bow.

It is assumed that the 1.8 minutes prior warning would have provided sufficient advance notice to plan and execute evasive maneuvers that may have lessened the severity of the grounding or averted it entirely. Halting the forward momentum of the vessel would not have been possible as this would require around 1,300 meters, with the vessel moving at 16 knots, and this distance was not available.[34] However, slowing the vessel combined with executing a hard turn to starboard upon receiving the warning commencing approximately 800 meters prior to Scole Rocks would have significantly reduced the damage incurred in the event of grounding such that the vessel may have remained afloat and lives may have been saved. Indeed, and even hypothetical, the accident may not have happened at all.

11.2.7 FLS Potential to Avert Disaster

Vessel groundings and collisions with HtoN can and do cause untold suffering, loss of life and property, and often result in devastating environmental damage. Since the *Costa Concordia* accident in 2012, many ships have continued to run aground primarily under circumstances where situational awareness was lacking by watchstanders of conditions surrounding the vessel below the waterline, oftentimes as their vessels were making good yet deviating from the planned track. The names of these ships are many, and a partial list includes: *Coastal Isle* (2012), *Tundra* (2012), USS *Guardian* (2013), *Ovit* (2013), *Maersk Garonne* (2015), *Hamburg* (2015), *Nova Cura* (2016), *Roger Blough* (2016), *Molly Manx* (2016), USS *Antietam* (2017), *Kea Trader* (2017), *Universal Durban* (2017), *Marbella* (2017), *Celtica Hav* (2018) *Priscilla* (2018), *Leda Maersk* (2018), *Wakashio* (2020), *Kaami* (2020), *Ever Given* (2020), *Beaumaiden* (2021), USS *Connecticut* (2021), *Ever Forward* (2022) and *Grande Senegal* (2023).

Costs incurred as a result of groundings must include property and liabilities arising from the accident itself as well as the cleanup. Such costs escalate as vessels increase in size and cargo capacity, especially for cargos that include toxic chemicals. Larger passenger vessels with greater carrying capacity create even greater risk. The effects of such accidents are amplified in areas that are remote where search and rescue and salvage efforts are problematic at best.

FLS can provide a means to directly detect bottom contours threatening to rise up to meet the hull, HtoN that are not present on navigation charts and can supplement existing methods to effect greater urgency to adverse circumstances and heighten situational awareness. The technical capabilities exist to make a substantial difference now. The establishment of carriage requirements as well as fusion into integrated navigation systems still needs to take place. The costs involved are minimal, especially in light of the consequences of not having FLS available for navigational use. The salvage cost of *Costa Concordia* alone amounted to over $2 billion US.[35] This amount by itself would cover the cost to purchase and install FLS equipment on 75% of the world's merchant fleet greater than 25,000 gross tons.[36] These same comments are just as applicable today as they were in 2014. Very little has changed.

11.2.8 Acknowledgements

Some of the results and parts of the investigations presented were performed under the European Interreg IVb-project – ACCSEAS Accessibility for Shipping, Efficiency Advantages and Sustainability. Portions of this paper were originally published by Michael Baldauf and R. Glenn Wright at the 10th International Symposium on Integrated Ship's Information Systems (ISIS 2014), German Inst. of Navigation, Deutsche Gesellschaft für Ortung und Navigation e. V. (DGON), Hamburg, Germany, 5 September 2014.

11.3 eCelestial Navigation System (eCNS)[2]

This paragraph describes a modern implementation of our greatest legacy method by using the sun, moon and stars to guide ship navigation at sea, and across inland bodies of water. It can also supplement pilotage techniques when physical aids to navigation are not visible or are unavailable, and when satellite methods such as Global Navigation Satellite System (GNSS) are stymied by degradation, denial of service attacks and spoofing events. Technology has progressed to the point where the development of an automated electronic Celestial Navigation System (eCNS) suitable for use by vessels of all types is entirely feasible. Conventionally manned, unmanned and autonomous vehicles can benefit from an independent source of positioning information for primary navigation and as a reference for Inertial Navigation System (INS) alignment.

The approach taken in original and ongoing research performed by the author uses high resolution, commercially available passive sensors responsive to visible, far-red and near-infrared light capable of imaging celestial objects, including the sun, moon and stars as well as manmade satellites during both daylight and nighttime conditions. Multi-megapixel cameras combined with machine learning and deep-learning artificial intelligence (AI) techniques make possible the continuous measurement of large numbers of sky object altitudes approaching one-arcsecond accuracy that can provide positioning to 31 meters at the Earth's surface. These sensors reside on a gyro-stabilized platform supplemented with software camera stabilization to provide isolation from vessel movement in moderate seas – all contained within a ruggedized waterproof housing topside approximately 0.75 cubic meters in volume and weighing less than 38 pounds.

11.3.1 eCNS Technical Specifications

A wide field of view of 360° horizontally and 180° vertically is possible with precision measurement maintained between 15° and 75° from the horizon ensuring system resilience across a wide range of weather conditions that may partially obscure the sky. Sensor power is provided and sensor data routed to a below-deck controller consisting of a power supply, computer and computer interfaces requiring a small amount of space (3U, or less) in a computer rack. Celestial object and satellite information are contained within databases that facilitate precise spatial-domain identification of individual objects as well as the making of multi-star, Sun and Sun-Moon fixes, and positioning based upon the orbital paths of observed satellites. An artificial intelligence-based automated process utilizing data fusion from multiple sensors establishes an artificial horizon and assists in sky object identification during obscured meteorological conditions and guides continuous observations and positioning updates at intervals of every few seconds throughout a voyage. This makes possible the identification of spatial and time-domain relationships to discern satellite behavior from those of aircraft, birds and other environmental features. Further, the design of the system readily detects and identifies eCNS degradation due to natural phenomena and distinguishes it from nefarious influences effects that may result from third-party actions. A capability also exists to capture and forward for remote human review one or more small snippets of sensor data via a low bandwidth communication channel as may be useful for off-vessel performance monitoring.

Design models, as well as prototypes of critical hardware components provide the means to assess system trades that include size, weight, power consumption, cost and projected system performance. In addition, prototype eCNS hardware mounted on an autonomous maritime vessel testbed research vessel evaluates system utility, installation options and limitations, and initial performance capabilities while underway.

11.3.2 eCNS Problem Identification

An investigation was undertaken into three distinct aspects of the celestial navigation problem. The first relates to the identification and specification of sensors and the design of an overall system architecture through which the eCNS perceives the physical environment external to the vessel as well as internal vessel capabilities to fulfill navigation requirements. This included how sensor data and the physical and environmental characteristics they represent are understood and the fusion of these data through which a comprehensive near-real-time (within fractions of a second) assessment of positioning awareness was created. Commercial-off-the-shelf hardware is used to create the complete eCNS. Mounted topside at a location with an unobstructed view of the sky in all quadrants is the ruggedized waterproof housing with a transparent dome attached to a stabilized platform. Within the housing are installed independent high-resolution visible light and infrared cameras with wide-angle optical lenses suitable for full sky coverage for each camera. Also contained within the housing are cooling and heating apparatus sufficient to prevent condensation from occurring on the inside of the housing and on the transparent dome. The platform is stabilized in two axes to compensate for ship's motion in terms of roll and pitch. Additional hardware is associated with the computer that forms the heart of eCNS computational capabilities and the stabilization controller that isolates the eCNS camera/optics housing from ship's movements.

Second, methods and processes used in celestial navigation are examined across many individual situations and operational scenarios considering reasoning with uncertainty and potential sensor degradation. A key element to ensure success was the use of existing commercial-off-the-shelf products to the greatest extent possible and not have to re-invent already proven technology in developing the eCNS architecture. The line of research dictated the use of numerous algorithms to perform celestial navigation in the maritime domain in terms of maritime piloting functions to provide data inputs through an automated process. The same issue applies to satellite databases, where reusing existing software provides significant cost and time savings and reduces overall program risk. The software chosen for this application contains details on more than 4,800 satellites, including over 4,000 Low Earth Orbit (LEO) satellites currently orbiting Earth and provides orbit information including apogee, perigee, inclination, period and other operational details.

The capability to acquire and process visible and far red/near-infrared camera imagery to derive position information is an essential aspect of this research. This is being accomplished using software functions to sample video images at a standard rate, detect prominent image contents (points of light) likely to represent navigational bodies (Sun, Moon, stars and satellites) and store these as data objects for subsequent use in a two-stage effort for object identification. Histograms of sky body images assist in the identification of individual observed object spectral contents through comparison with known objects. Supplementing this software are machine learning algorithms that identify persistent objects (i.e., not satellites) based on their position in the sky along with their proximity to other objects. The results of these two steps generate a probability of identification as a particular object, which is then assessed to either reinforce this conclusion or to choose another, more likely object identification. Such methods are essential under partially obscured skies when only portions of constellations may be visible. Another aspect of this effort is the use of machine learning techniques to establish an artificial horizon during obscured conditions through training with known horizon characteristics under unobscured conditions for the eCNS installation. This is an essential element in establishing a pixel-width precision baseline by which a pixel-by-pixel correlation may be made between the position of the camera's sensor element and the position of an object in the sky that is essential to achieve highly accurate object altitude measurements that is far beyond the capability of human observers.

Another aspect of this development effort is to again use machine learning to detect and differentiate between relatively persistent objects, such as stars and planets and moving objects in the sky that include satellites and aircraft. This is accomplished through frame-by-frame comparison to reveal moving objects, and histogram analysis of moving object images that detect aircraft strobe and navigation light characteristics that are not present in satellites. A third aspect includes detection of the distinct and unique light characteristics of anomalies and behavior that may represent sensor degradation due to natural conditions and those resulting from the nefarious acts of adversaries.[37]

11.3.3 Present Development Status

Ongoing efforts continue with refinement of AI-based software and conventional algorithms that implement eCNS user interface features. Also included is further development of interfaces with celestial almanac and satellite databases, provisioning of INS positioning and timing data, interacting with other onboard systems as appropriate, and system anomaly handling as may be needed. Future presentation of case studies and validated test results is anticipated.

11.4 Virtual Aids to Navigation (VAtoN)

This paragraph describes new technology for implementing truly Virtual Aids to Navigation (VAtoN) that have no need to rely on the presence of VHF radio physical infrastructure within line of sight to establish the presence of navigation aids through Automatic Identification System (AIS) technology. Rather, VAtoN exist on the navigation chart at designated positions defined by waypoints that are georeferenced (or anchored, so to speak) to local bottom topography. They require no maintenance other than periodic hydrographic surveys to ensure the bottom depths and contours on which their positions are based remain accurate so that a proper watch may be maintained – a prudent task regardless of whether physical or virtual aids to navigation are deployed.

11.4.1 VAtoN Need and Application

Earlier in Chapter 3, Navigation, there was mention of the use of satellite navigation where in practice almost universal reliance upon this single source of PNT data is unwisely made. GNSS sensor data has been integrated and is shared throughout all bridge systems and across sensor data bus interconnections throughout the vessel. However, the susceptibility of GNSS to spoofing and denial of service attacks and the subtle degradation in positioning accuracy and reliability that may ensue may catch watchstanders unaware and lead to loss of situational awareness and even disaster. Attempts to mitigate and overcome these frailties are many and include the development of enhanced encryption, anti-spoofing techniques and other methods to overcome denial-of-service attacks. However, adversaries who are intent on malicious behavior continue to make advances just as fast as countermeasures can be developed.

11.4.2 VAtoN Development

A means was developed to maintain situational awareness sufficient to safely navigate vessels without the need to rely on GNSS technology whereby crewmembers and the vessels on which they serve may again be self-reliant and not at the mercy of external entities for the real-time information they need to navigate and operate. This is envisioned as involving active navigation systems such as Radar, AIS, Sonar and physical AtoN; as well as a passive navigation system that facilitates precise positioning by adapting traditional methods for pilotage to present day and evolving environmental sensing capabilities that exist within the vessel itself. This approach also takes advantage of high-resolution hydrographic vector data products provided by unmanned survey vehicles, crowd-sourced data produced by participating surface vessels, and by already available assets established by national hydrographic agencies and other available civilian resources. The underlying concept is to create a navigation system that considers a wide range of environmental and vessel factors to update watchstanders INS guidance information. The approach acknowledges the obtaining and fusion of data from multiple, independent onboard sensor systems that include:

- Traditional single beam echosounder to determine real-time depth and bottom contour information for comparison with legacy charted depths contained within ECDIS to determine probable location.
- Side-scan Sonar for seabed feature detection and identification as physical aids to navigation for pilotage on the sea bed and to precisely identify known positions.
- Electronic Celestial Navigation System (eCNS) for day and night determination of position fixes from astronomical objects, including stars, the sun, moon, planets and satellites.

- Visual piloting using visible light and infrared cameras to detect and identify physical aids to navigation present at the sea surface, including buoys, lights and ranges plus landmarks and terrain features.
- Virtual AtoN (VAtoN) georeferenced to the seabed environment as waypoints along unmarked routes.

11.4.3 Sensor Utilization

The basic single-beam echosounders that exist onboard vessels by design and regulation through mandated carriage requirements provide data needed to help determine the depth of water that exists between the bottom of the vessel and the seabed and is vital to ensure it does not run aground. Sailors have developed the means to navigate using bottom contours and, aided by computer automation, this capability has become much more accurate and is limited only by the resolution and accuracy of hydrographic surveys and their results codified in modern navigation charts. The use of numerical data searches of ECDIS database contents to detect patterns within vector navigation chart depth soundings corresponding with continuous live Sonar measurements of depth, change of depth and rate of change of depth patterns can greatly aid in identifying actual present vessel position.

Supplementing the single-beam echosounder with side-scan Sonar helps detect, recognize and positively identify seabed features such as significant natural and/or manmade landmarks along with bottom consistency (e.g., rock, sand and mud). All such physical features and landmarks are themselves georeferenced to the physical environment enabling them to function as AtoN, and waypoints that are also georeferenced to the physical environment enabling them to function as Virtual AtoN; both of which are verifiable as watching properly by having the proper characteristics and being in the correct position. This approach extends traditional pilotage that takes advantage of hills and bluffs, water and transmission towers, river mouths and other natural and manmade landscape features to determine location; to extend below the waterline where known boulders, reefs, fissures, wrecks and possibly manmade aids to navigation disguised as natural features can mark precise locations capable of being used as a position reference for inertial navigation. The task of AtoN detection, recognition and positive identification is one that is perfectly suited for machine learning and has been demonstrated by the author and others as being practical. Similar to the echosounder, side-scan Sonar (available as a built-in feature on many low-cost commercial echosounders and fish finders) can assist when position error crosses a certain threshold and when the vessel is believed to be in relatively close proximity to a known undersea feature.

Additional considerations include the effects of tides, winds, sea state and other external factors affecting a voyage; combined with their effects on performance-based on vessel motion in terms of heave, sway, surge, roll, pitch and yaw. Also considered can be internal sensors and external meteorological and oceanographic (METOC) sensor data content and availability to derive error correction vectors that compensate for progressive error exhibited by INS resulting from a lack of positional references through GNSS or known physical location. This capability utilizes software developed through machine-learning techniques to model natural forces and tailor their effects on performance to the unique characteristics of a specific vessel. Deep-learning AI can also be used to discern unknown factors, trends and events present in the data that may also influence vessel performance.

The combination of such techniques can provide a means for a navigation system to constantly learn new things about the environment and the vessel itself and to improve its own operational capabilities on an ongoing basis. All such methods and capabilities listed above are possible today using existing computing and AI technology. Each method is subject to its own limitations and accuracy characteristics. However, the fusion of multiple sources of complementary data can reinforce conclusions reached regarding positioning and achieve accuracies greater than those of individual sources.

An autonomous vehicle would not have the constraints of human senses and attributes but, without a human in the loop, would lack the knowledge and decision-making necessary to properly execute safe passage. Machine learning and optical character recognition can be used to train an automaton to recognize basic shapes and colors of standard buoys and other AtoN, and to read identifying numbers and letters that would enable positive identification of individual navigation aids. An automated process is used to correlate with ECDIS and the Light List along with performing cross-checking with Radar and GNSS indications to verify that the detected aid is watching properly in the correct position and exhibiting proper characteristics. However, less than ideal orientation of buoy numbers and letters, fouling by birds, low light conditions and other natural and man-made phenomena can drastically obscure, alter physical and light characteristics and reduce the certainty of what is being seen. Other charted AtoN, such as lighthouses, piers, water towers, building and landscape features are much less standardized, but a mariner would routinely detect and identify can be much more challenging or outright impossible for automated processes.

11.4.4 VAtoN Experiments

Figure 11.5 illustrates an experiment performed to detect and identify AtoN to assist in the automated navigation of a vessel across a setting that includes different aspects of open sea, harbor and inland navigation.[38] This was accomplished entirely without the benefit of GNSS. The methodology incorporates the use of machine learning, reinforcement learning and deep learning AI techniques to create image, data and information analytics for multiple sensor modalities for a cohesive approach to Maritime Autonomous Surface

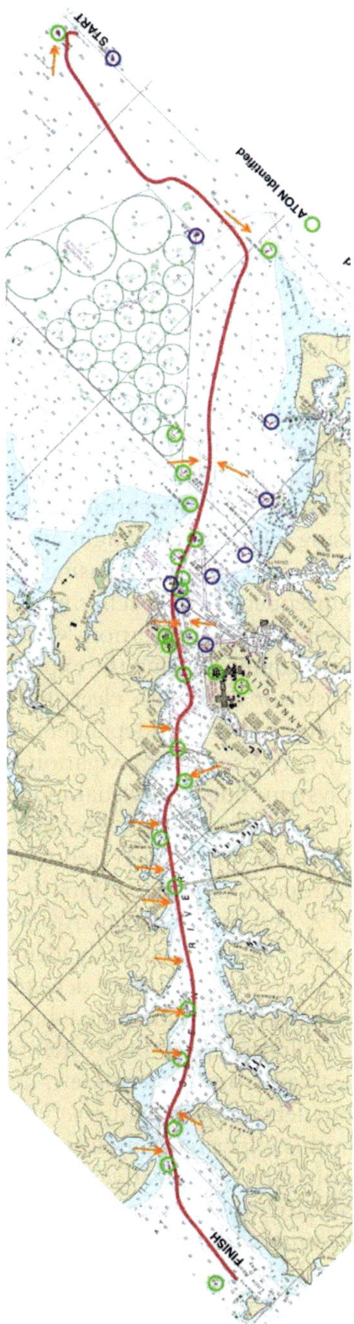

Figure 11.5 Aid to Navigation (AtoN) Detection and Identification Experiment Route.

Ship (MASS) navigation. However, issues associated with the use of AI in safety-critical MASS applications cannot be examined in uncertain, unstructured and dynamic environments which lack mathematical frameworks to verify the correctness of their solutions and validate the appropriateness of their actions.

The location of this experiment was the Chesapeake Bay and Severn River near Annapolis, Maryland in the United States, where vessel traffic includes large oceangoing ships, tugs, barges and many small recreational boats. Charted landmarks abound that are useful as AtoN, including buildings, domes and natural terrain, including 27 buoys and fixed aids, 2 bridges and 4 buildings selected for this experiment. The transit route was approximately 11 nautical miles in length with an average depth of 31 meters (east) and 5 meters (west).

A ten-meter research vessel shown in Figure 10.1 configured as a maritime autonomous system test bed was used to perform this experiment. Sensor data acquisition (e.g., compass, Radar, AIS, echosounder) was accomplished in real time during the transit on the navigation computer with subsequent fusion during post-processing. Daylight, low light and infrared imaging data along with Radar and FLS data were captured using video data recorder and NMEA-2000 bus, with subsequent fusion achieved during post-processing. Automated AtoN detection and identification were performed live and recorded on the navigation computer during the experiment. Post-processing correlated and verified all sensor data with electronic chart tracking information and GNSS positioning acquired independent of the experiment as a secondary check. For safety reasons, vessel navigation was performed uncoupled from the navigation system, but with visual indications provided at all times to the helm and followed throughout the voyage.

Updated results from this experiment were reported at the International Conference on Maritime Autonomous Surface Ships (ICMASS) in Trondheim, Norway, in November 2019.[39] Of the 27 marine AtoN (buoys) present along the route, all were detected and 18 were positively identified. In addition, two bridges and four buildings along the route were detected and positively identified. Nine buoys were detected but not identified due to distances being too great and beyond sensor resolution, poor buoy orientation with identifying characteristics positioned away from camera view, and fouling preventing optical recognition of buoy identification features.

Conclusions taken from this experiment indicate that smart sensors combined with well-trained software can positively identify manmade critical AtoN and provide adequate situational awareness in an operational environment under defined circumstances for onboard computers. However, remote operators may be hampered by bandwidth and data latency issues. Further conclusions indicated that top-level (Captain) decision-making capabilities must further mature before it can be determined whether overall system resilience in managing distributed smart sensors can be assured.

Pilotage navigation of a vessel through restricted waters has traditionally involved the use of terrain features, landmarks and other AtoN such as buoys visible to watchstanders marking proper channels and hazards to navigation. Awareness of the depths of the waters below the keel to ensure safety of navigation is also an essential part of pilotage first achieved in antiquity using the leaded line, wire drags and subsequently through the use of the single-beam echosounder beginning in the 1930's.[40] In 1974, under SOLAS, echosounders were specified under IMO mandatory carriage requirements for larger vessels, and they are now standard equipment found on just about all vessel types and sizes larger than a rowboat. The relatively recent introduction over the last 50 years of multibeam and side-scan Sonar systems has greatly advanced hydrographic survey practices and vastly improved the accuracy and resolution of soundings data on navigation charts. These inventions, along with FLS, now enable individual vessels to obtain highly accurate visibility into the seabed configuration not only below the keel but ahead of the vessel's path to instantly warn of shoals, reefs and other hazards to navigation and allow sufficient time to take evasive action prior to running aground.

One method of piloting using an echosounder is the use of depth contour following to determine when the vessel enters deeper or shallower water. This is accomplished using waypoints to establish desired entry and/or exit points along a depth contour that may be used to roughly establish position when anticipating crossing a bank or upon approaching a seacoast featuring gradually shallowing depths. Another method is to follow a contour at a specific depth until reaching a prominent landmark or AtoN that may lead to the destination.

Enhanced Sonar capabilities, to directly assess bottom configurations and conditions in real time, provide new and redundant precision navigation and vision capabilities to provide terrain tracking capabilities and high-resolution imaging of prominent seabed features in the same way as surface landmarks to aid navigation in ways never before possible.[41] Such capabilities are essential for Unmanned Surface Vehicle (USV) navigation without GNSS but can also apply to conventional vessels in the future. Navigation Sonar with forward-looking capabilities can provide high-resolution bathymetry that may be compared to electronic navigation charts (ENCs) displayed on ECDIS for backup navigation, detection of hazards to navigation and obstacles, and avoidance of large marine mammals.[42] These data are sufficient to crowdsource navigation chart bathymetry.[43]

Figure 11.6 illustrates the combined use of visual AtoN detection and identification with navigation by depth contour tracking. Originally developed for Arctic navigation in poorly charted and uncharted regions, the principles involve the correlation of three aspects of bathymetry to determine positioning: the depth soundings themselves, changes in depth soundings along the planned course, and the rate of change of depth soundings as they occur.[44]

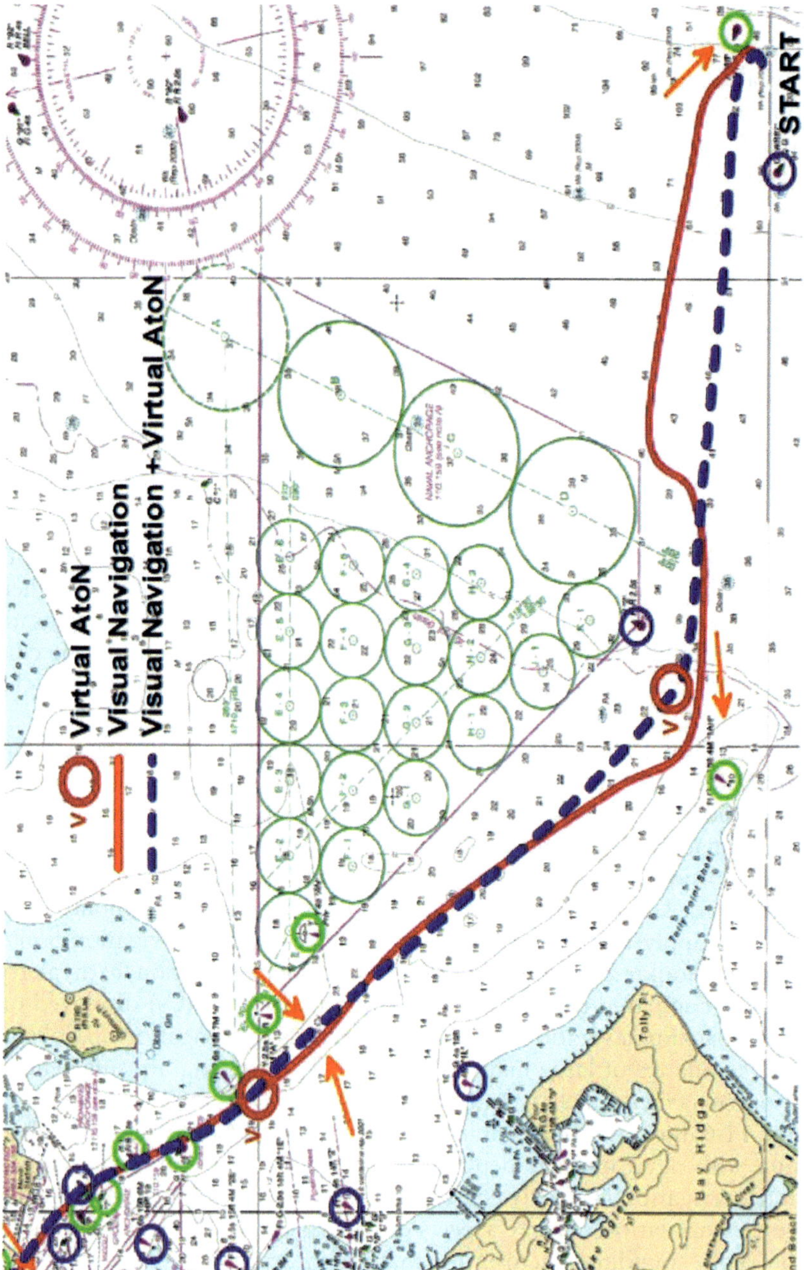

Figure 11.6 AtoN Detection and Identification Combined with Depth Contour Following.

This approach formed the basis to implement truly Virtual AtoN (VAtoN) that do not require a corresponding VHF radio transmitter but are georeferenced to the physical seabed environment as waypoints. Course and heading information developed by the navigation computer using multi-sensor data fusion from cameras, Radar, Sonar and other sensors are provided directly to the helmsman and autopilot to execute the voyage plan.

The findings of this experiment note the significant contribution of depth contour following and the use of VAtoN depicted as red circles to enhance adherence to the planned route over the initial portion of the experiment course where physical AtoN in the form of buoys are sparse with great distances between buoys that are essentially beyond camera range. Further along the course shown in Figure 11.6 where the channel was well marked with physical AtoN, the use of depth contour following using VAtoN was generally effective in maintaining a proper course within the channel. Note that the illustrations provided are shown using raster chart depictions for clarity. This experiment was performed using vector charts and supplemented with live soundings obtained using the FarSounder ARGOS 350 forward-looking navigation Sonar (FLS) to provide more accurate data in much greater detail than is available using vector charts.

11.5 Quantum Sensing

Welcome to the weird, wonderful world of the quantum domain, where seemingly the laws of physics that govern the behavior of familiar objects give way to the principles of quantum mechanics that describe interactions at the atomic and subatomic levels. I'll go into this in more detail a little later. But first, you might wonder what quantum sensing has to do with ship sensors. Quantum sensors can measure time, dynamics (i.e., forces, acceleration and rotation) and fields (i.e., gravitational, electromagnetic and mechanical) with much greater precision and stability than any presently available technology. These properties have boundless potential for improving existing ship sensors and sensor types, and inventing previously unimaginable sensing technologies that can provide new solutions to current problems and difficulties.

11.5.1 Quantum Ship Sensor Applications

There are many different areas on ships and within ship functions where quantum sensing can make significant improvements in existing and future sensing applications. Examples include:

- **Ship Navigation:**[45] A quantum sensor with potential applications in GNSS-free navigation, developed at Imperial College London, has been tested in collaboration with the Royal Navy. This sensor represents a new type of accelerometer that displays quantum properties at ultracold temperatures. By combining changes in an object's velocity over time with rotation measurements and the initial position of the object, the current location of a ship can be calculated with great precision.

- **Temperature Sensing:**[46] Quantum dots, which are three-dimensional semiconductor nanoparticles ranging in size from 2 to 10 nm, feature fine-tunable optical characteristics and can be highly sensitive to environmental conditions such as temperature. These specific properties make them useful for a wide range of applications, including optical and opto-electronic devices for temperature sensing for machinery and machinery spaces, cargo and environmental monitoring.
- **Imaging:**[47] Quantum sensors can be combined with imaging devices to perform new types of ultra-high resolution imaging capabilities that far surpass those of present-day camera and video, Radar, Sonar and Lidar technologies. This can include detecting and building 3D models of stealth or "invisible" objects by sensing gravitational changes in the local area. These devices can include a probe that scans the position of the sensor, an array of sensors, or a beam of electromagnetic waves prepared in a quantum state to perform high-resolution macroscopic (i.e., large-scale) imaging with unprecedented sensitivity. Also possible are new features that can see around corners and through walls without any adverse effects on materials or living creatures.
- **Non-Destructive and Non-invasive Testing:**[48] Combining groups of atoms and a web of lasers, scientists can monitor the fluctuations in individual atoms within this web. This can culminate in large amounts of accurate data that can be used to detect metal fatigue, failures and pending failures, and other changes within rotating machinery such as propeller shafts, mounts for deck and hull equipment, and other mechanical and electro-mechanical devices.
- **Quantum Twin Duplication:**[49] The quantum property of entanglement can be incorporated into sensors where two particles within the apparatus would be entangled or have interdependent quantum states. This can have the effect of being able to duplicate events taking place at one location for viewing or taking action at another, such as between a ship and a shore location. Entanglement can occur theoretically without regard to distance.
- **Sensor Security:**[50] Quantum metrology leverages quantum entanglement to surpass existing physical limitations and achieve great measurement precision. Quantum states also have a non-deterministic nature leading to uncertainties in their characterization and can be employed in quantum cryptography protocols to provide information and data security.

These are just a few of the many areas of research that can be applied directly to the maritime environment and, specifically, to ships and unmanned vehicles or all types and kinds.

11.5.2 How Quantum Sensing Works

Quantum sensors use the principles of quantum mechanics to measure physical properties at the atomic and subatomic levels, which behave quite differently than the normal physical objects we deal with every day. In the quantum domain something can be either one thing or another at the same time – until you look at it. Then it becomes one thing. Many principles of quantum mechanics achieve capabilities that do not necessarily make sense using conventional physics. Let me explain.

The "quantum bit" (qubit) is very much like the "bit" of conventional computing that stores information as either a 1 or 0. This is where the similarity ends since a qubit can also be in a combination (or *superposition*) of both states or somewhere in between at the same time. However, when you observe the quantum bit then it becomes a 1 or 0. This is sort of like the cat of someone named Schrödinger who put it in a sealed box (metaphorically, of course) where it could either be dead or alive (or both?) at the same time.[51] However, when you peek inside the box, then you know one way or the other. Now that you understand this basic principle you have graduated to the level of quantum engineer (third class) and are now ready for the next level.

- Sensors using quantum mechanics can be linked together (or *entangled*) in such a way that they share the same fate. This allows quantum sensors to achieve even higher sensitivity and to operate independently yet produce the same results regardless of distance.
- Quantum sensors can observe phenomena by *tunneling* particles with a wave function defined over all space that can go through barriers impossible for classical objects to pass through. This allows quantum sensors to measure very small changes in electric fields and distinguish between physical properties with great precision.
- The ability to reduce unwanted noise caused by photons popping in and out of existence in empty space is referred to as quantum *squeezing*. This occurs by reducing uncertainty in one property at the expense of increasing the uncertainty in another property, allowing quantum sensors to achieve even higher precision.

These are just a few of the principles of quantum mechanics that are used in quantum sensors. As quantum sensing technology continues to develop, it is likely that even more ways will be discovered to use quantum mechanics and to achieve even higher sensitivity and precision in sensors.

Congratulations! Understanding these basic principles advances you to quantum engineer (second class). However, this is as far as you can go because the rest of the curriculum has yet to be invented.

11.5.3 Future Directions of Quantum Sensing

Quantum technology is expected to change and improve secure communication and information processing, improve the sensitivity of detectors and sensors, provide new functional materials for future electronic devices and change the way we approach computing.[52] In theory, quantum sensors are able to achieve much higher sensitivity and precision than conventional sensors, making them well-suited for a wide range of applications including navigation, process and environmental monitoring, and even space exploration. There are many areas in quantum mechanics that are yet to be explained. Albert Einstein called quantum entanglement "spooky action at a distance", where two particles potentially separated by light-years great distances but connected through a mysterious and inexplicable link. At present, there are about 30 different models still trying to explain it.

Notes

1 In this paragraph (11.2) INS is used for Integrated Navigation System, NOT Inertial Navigation System.
2 In this paragraph (11.3) INS is used for Inertial Navigation System, NOT Integrated Navigation System.

References

1 Drone market outlook in 2023: Industry growth trends, market stats and forecast. Insider Intelligence. 7 January 2023. www.insiderintelligence.com/insights/drone-industry-analysis-market-trends-growth-forecasts/

2 Matthew Spaniol. (2020). Drones on Ships. Periscope. 10.13140/RG.2.2.18715.28969. www.researchgate.net/publication/347521511_Drones_on_ships

3 Aishwarya Lakshmi. Maersk to Use Drones to Resupply Its Fleet of Tankers. 6 June 2016. MarineLink. www.marinelink.com/news/resupply-tankers-UAS410695.aspx

4 Gary. How Drones Are Changing the Maritime Industry. 18 May 2016. Ship Technology. www.ship-technology.com/features/featurehow-drones-are-changing-the-maritime-industry-4865807/

5 Wilhelmsen and Airbus Trial Drone Deliveries to Singapore Anchorage. 16 March 2019. The Maritime Executive. https://maritime-executive.com/article/wilhelmsen-and-airbus-trial-drone-deliveries-to-singapore-anchorage

6 Carrier USS Ford Tests Out Air Cargo Drone. 8 August 2021. The Maritime Executive. www.maritime-executive.com/article/video-carrier-uss-ford-tests-out-air-cargo-drone

7 www.cruisejunkie.com/Overboard.html

8 Marine Accident Recommendations and Statistics, 2015–2022. Marine Accident Investigation Branch (MAIB), United Kingdom.

9 A.S. Selmy, (2016). The Need of Man Overboard (MOB) Detecting and Tracking System Descriptive Analyses. Conference: The 9th China International Rescue & Salvage Conference, China.
10 Orhan Gönel and Ismail Çiçek. ÖNEL Statistical Analysis of Man Overboard (MOB) Incidents. December 2020. www.researchgate.net/publication/348266442_STATISTICAL_ANALYSIS_OF_MAN_OVER_BOARD_MOB_INCIDENTS
11 Martin Manaranche. Royal Navy Tests Drones in Man Overboard Trials. Naval News, 06 July 2021. www.navalnews.com/naval-news/2021/07/royal-navy-tests-drones-in-man-overboard-trials/
12 Donato Cafarelli, et al. MOBDrone: A Drone Video Dataset for Man OverBoard Rescue. 15 March 2022. 22st International Conference on Image Analysis and Processing. ICIAP 2021.arXiv:2203.07973. https://doi.org/10.48550/arXiv.2203.0797
13 Mike Ball. Antarctic Supply Ship Uses Drone to Navigate Sea Ice. Unmanned Systems Technology , 23 December 2015. www.unmannedsystemstechnology.com/2015/12/antarctic-supply-ship-uses-drone-to-navigate-sea-ice/
14 Royal Dutch Shell Scope of Work Narrative. Anchor Retrieval Project (Location: Beaufort and Chukchi Seas). Fairweather, LLC. Anchorage, Alaska. USA. 1 August 2023.
15 Drone Guided Ship through Arctic Sea Ice. 16 November 2016. iHLS. https://i-hls.com/archives/72848
16 Matt McFarland. Ship relies on drone to avoid ice blocks in Arctic waters. CNN Business. 5 November 2016. https://money.cnn.com/2016/11/05/technology/arctic-drone-ship-navigate/index.html
17 Drone Survey of ONE Apus Container Collapse. 16 December 2020. The Maritime Executive. www.maritime-executive.com/article/video-survey-of-one-apus-container-collapse-made-public-by-wk-webster
18 Gary. How drones are changing the maritime industry. 18 May 2016. Ship Technology. www.ship-technology.com/features/featurehow-drones-are-changing-the-maritime-industry-4865807/
19 International Regulations for Preventing Collisions at Sea 1972 (COLREGS), Rule 5: Look-out.
20 Standards of Training, Certification and Watchkeeping (STCW) Convention and Code, Chapter VIII, section A-VIII/2, part 3 – Watchkeeping at Sea.
21 Samuel Halpern, An Objective Forensic Analysis of the Collision Between Stockholm and Andrea Doria, www.titanicology.com/AndreaDoria/Stockholm-Andrea_Doria_Collision_Analysis.pdf
22 Resolution A.918(22), IMO Standard Marine Communication Phrases, p. 67, B1/1.3.
23 United States Patent 5,390,152; Boucher, et al.; 14 February 1995.
24 United States Patent 5,675,552; Hicks, et al.; 7 October 1997.
25 United States Patent 7,173,879; Zimmerman, et al.; 6 February 2007.
26 United States Patent 8,717,847; Blake, 6 May 2014.
27 IMO Resolution MSC.232(82): 2006, Adoption of the Revised Performance Standards for Electronic Chart Display and Information Systems (ECDIS).
28 IMO Resolution A.1021(26), Code on Alerts and Indicators, 2009.

29 Guidance Notes on Ergonomic Design of Navigation Bridges, October 2003, American Bureau of Shipping,
30 Operation Manual, FarSounder-1000 Navigation Sonar, F31552 (Rev. 2.9.0) 2014, FarSounder, Inc.
31 Marine Casualties Investigative Body, Cruise Ship Costa Concordia, Marine casualty on 13 January 2012, Report on the safety technical investigation, pp. 3, 13, 14; Ministry of Infrastructures and Transports (Italy).
32 Isle del Giglio, www.cruisersforum.com/forums/f134/chart-accuracy-and-detail-comparison-74805.html
33 Marine Casualties Investigative Body, Cruise Ship Costa Concordia, Marine casualty on 13 January 2012, Preliminary Report, Slide 14, 18 May 2012, presented at MSC90, London.
34 Marine Casualties Investigative Body, Cruise Ship Costa Concordia, Marine casualty on 13 January 2012, Report on the safety technical investigation, p. 50, para. 2; Ministry of Infrastructures and Transports (Italy).
35 Telegraph 2014, Costa Concordia salvage: the £1.2bn price tag of the disaster; The Telegraph, 14 July 2014, London, www.telegraph.co.uk/news/worldnews/europe/italy/10966308/Costa-Concordia-salvage-the-1.2bn-price-tag-of-the-disaster.html
36 Equasis Statistics; The World Merchant Fleet in 2011, Table 1, p. 6.
37 GMATEK, Inc. In-Stride Detection of Sensor Degradation. Final Report. 28 April 2022. Contract N68335-20-G- 2005. Principal Investigator: R. Glenn Wright, Ph.D.
38 R. Glenn Wright, (2019). Intelligent autonomous ship navigation using multi-sensor modalities. *International Journal on Marine Navigation and Safety of Sea Transportation*, 13(3). September 2019. DOI: 10.12716/1001.13.03.03 www.transnav.eu
39 R. Glenn Wright. Enhanced MASS situational awareness using virtual navigation aids. International Conference on Maritime Autonomous Surface Ships (ICMASS) in Trondheim, Norway in November 2019.
40 NOAA. History of Hydrographic Surveying. Office of Coast Survey. National Oceanographic and Atmospheric Administration. www.nauticalcharts.noaa.gov/learn/history-of-hydrographic-surveying.html
41 Wright R. Glenn and Baldauf M. (2016). Correlation of Virtual Aids to Navigation to the Physical Environment. TransNav, The International Journal on Marine Navigation and Safety of Sea Transportation, 10(2), 287–299. DOI: 10.12716/1001.10.02.11
42 R. Glenn Wright and Ian Russell (2017). Navigation Sonar Use in Maritime Frontier Exploration. Soundings. Hydrographic Society of the UK. Summer 2017.
43 FarSounder 2018. FarSounder Joins NOAA as a Trusted Node. Press Release. 17 October 2018. www.farsounder.com/about/press_releases
44 R. Glenn Wright (2017). 4-dimensional Virtual Aids to Navigation for Arctic Maritime Transit. IEEE/MTS Oceans Conference. Anchorage, AK. Session AK.10. Marine Safety and Security with Diminishing Ice. September 2017.

45 Hayley Dunning. Quantum Sensor for a Future Navigation System Tested Aboard Royal Navy Ship. TechXplore. 26 May 2023. https://techxplore.com/news/2023-05-quantum-sensor-future-aboard-royal.html

46 Natalia Doskaliuk, Yuliana Lukan, and Yuriy Khalavka. Quantum Dots for Temperature Sensing. April 2023. DOI: 10.58332/scirad2023v2i2a01. www.researchgate.net/publication/370142487_Quantum_dots_for _temperature_sensing

47 Quantum Technology: Sensing and Imaging. Australian Army Research Centre. 2 September 2020. https://researchcentre.army.gov.au/library/land-power-forum/quantum-technology-sensing-and-imaging

48 Chen-Wei Liang, Ehsan Ahmad, Ertan Balaban, James Sexton, and Mohamed Missous. A Quantum Well Hall Effect Magnetovision System for Non-Destructive Testing. 19th World Conference on Non-Destructive Testing 2016. www.ndt.net/article/wcndt2016/papers/mo1f1.pdf

49 L. Pezzè. Entanglement-enhanced sensor networks. Nature Photonics 15, 74–76 (2021). https://doi.org/10.1038/s41566-020-00755-x

50 M.T. Rahim, A. Khan, U. Khalid. et al. (2023) Quantum secure metrology for network sensing-based applications. Science Report 13, 11630. https://doi.org/10.1038/s41598-023-38802-6

51 John Gribben. In Search of Schrödinger's Cat. Bantam Books. 1984. ISBN 978-0-553-34253-6 www.randomhousebooks.com

52 Tamara Dietrich. Curious about Quantum? 16 February 2023 https://new.nsf.gov/science-matters/curious-about-quantum

Index

9781032456218